THE GEOMETRY OF SPECIAL RELATIVITY

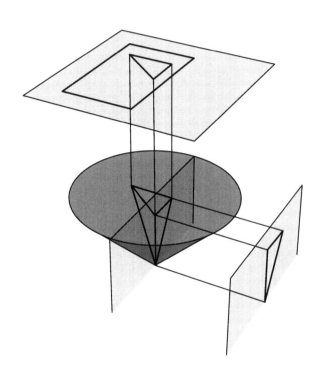

THE GEOMETRY OF SPECIAL RELATIVITY

TEVIAN DRAY

CRC Press
Taylor & Francis Group
Boca Raton London New York

CRC Press is an imprint of the
Taylor & Francis Group, an **informa** business
AN A K PETERS BOOK

Cover photo: Shukhov radio tower in Moscow, Russia. Designed by Vladimir Shukhov, it is a 138-meter-high hyperboloid structure.

CRC Press
Taylor & Francis Group
6000 Broken Sound Parkway NW, Suite 300
Boca Raton, FL 33487-2742

© 2012 by Taylor & Francis Group, LLC
CRC Press is an imprint of Taylor & Francis Group, an Informa business

No claim to original U.S. Government works

Version Date: 20120420

International Standard Book Number: 978-1-4665-1047-0 (Hardback)

This book contains information obtained from authentic and highly regarded sources. Reasonable efforts have been made to publish reliable data and information, but the author and publisher cannot assume responsibility for the validity of all materials or the consequences of their use. The authors and publishers have attempted to trace the copyright holders of all material reproduced in this publication and apologize to copyright holders if permission to publish in this form has not been obtained. If any copyright material has not been acknowledged please write and let us know so we may rectify in any future reprint.

Except as permitted under U.S. Copyright Law, no part of this book may be reprinted, reproduced, transmitted, or utilized in any form by any electronic, mechanical, or other means, now known or hereafter invented, including photocopying, microfilming, and recording, or in any information storage or retrieval system, without written permission from the publishers.

For permission to photocopy or use material electronically from this work, please access www.copyright.com (http://www.copyright.com/) or contact the Copyright Clearance Center, Inc. (CCC), 222 Rosewood Drive, Danvers, MA 01923, 978-750-8400. CCC is a not-for-profit organization that provides licenses and registration for a variety of users. For organizations that have been granted a photocopy license by the CCC, a separate system of payment has been arranged.

Trademark Notice: Product or corporate names may be trademarks or registered trademarks, and are used only for identification and explanation without intent to infringe.

Library of Congress Cataloging-in-Publication Data

Dray, Tevian.
  The geometry of special relativity / Tevian Dray.
      p. cm.
  Includes bibliographical references and index.
  ISBN 978-1-4665-1047-0 (hardcover : alk. paper)
  1. Special relativity (Physics) 2. Space and time--Mathematical models. I. Title.

QC173.65.D73 2012
530.11--dc23   2012010206

Visit the Taylor & Francis Web site at
http://www.taylorandfrancis.com

and the CRC Press Web site at
http://www.crcpress.com

Lorentz transformations are just hyperbolic rotations.

Contents

List of Figures and Tables — XI

Preface — XV

Acknowledgments — XVII

1 Introduction — 1
   1.1 Newton's Relativity — 1
   1.2 Einstein's Relativity — 2

2 The Physics of Special Relativity — 3
   2.1 Observers and Measurement — 3
   2.2 The Postulates of Special Relativity — 3
   2.3 Time Dilation and Length Contraction — 7
   2.4 Lorentz Transformations — 10
   2.5 Addition of Velocities — 11
   2.6 The Interval — 12

3 Circle Geometry — 13
   3.1 The Geometry of Trigonometry — 13
   3.2 Distance — 13
   3.3 Circle Trigonometry — 13
   3.4 Triangle Trigonometry — 15
   3.5 Rotations — 16
   3.6 Projections — 17
   3.7 Addition Formulas — 18

4	Hyperbola Geometry	19
	4.1 Hyperbolic Trigonometry	19
	4.2 Distance	20
	4.3 Hyperbola Trigonometry	20
	4.4 Triangle Trigonometry	22
	4.5 Rotations	23
	4.6 Projections	24
	4.7 Addition Formulas	24
5	The Geometry of Special Relativity	25
	5.1 The Surveyors	25
	5.2 Spacetime Diagrams	26
	5.3 Lorentz Transformations	27
	5.4 Space and Time	29
	5.5 Dot Product	30
6	Applications	37
	6.1 Drawing Spacetime Diagrams	37
	6.2 Addition of Velocities	38
	6.3 Length Contraction	39
	6.4 Time Dilation	41
	6.5 Doppler Shift	44
7	Problems I	47
	7.1 Practice	47
	7.2 The Getaway	51
	7.3 Angles Are Not Invariant	52
	7.4 Interstellar Travel	55
	7.5 Cosmic Rays	57
	7.6 Doppler Effect	59
8	Paradoxes	61
	8.1 Special Relativity Paradoxes	61
	8.2 The Pole and Barn Paradox	61
	8.3 The Twin Paradox	63
	8.4 Manhole Covers	65
9	Relativistic Mechanics	67
	9.1 Proper Time	67
	9.2 Velocity	68
	9.3 Conservation Laws	69

	9.4 Energy	71
	9.5 Useful Formulas	73
10	**Problems II**	**75**
	10.1 Mass Isn't Conserved	75
	10.2 Identical Particles	76
	10.3 Pion Decay I	77
	10.4 Mass and Energy	80
	10.5 Pion Decay II	81
11	**Relativistic Electromagnetism**	**83**
	11.1 Magnetism from Electricity	83
	11.2 Lorentz Transformations	86
	11.3 Vectors	89
	11.4 Tensors	91
	11.5 The Electromagnetic Field	92
	11.6 Maxwell's Equations	93
	11.7 The Unification of Special Relativity	96
12	**Problems III**	**97**
	12.1 Vanishing Fields	97
	12.2 Parallel and Perpendicular Fields	98
13	**Beyond Special Relativity**	**99**
	13.1 Problems with Special Relativity	99
	13.2 Tidal Effects	100
	13.3 Differential Geometry	102
	13.4 General Relativity	103
	13.5 Uniform Acceleration and Black Holes	105
14	**Hyperbolic Geometry**	**107**
	14.1 Non-Euclidean Geometry	107
	14.2 The Hyperboloid	108
	14.3 The Poincaré Disk	110
	14.4 The Klein Disk	113
	14.5 The Pseudosphere	114
15	**Calculus**	**119**
	15.1 Circle Trigonometry	119
	15.2 Hyperbolic Trigonometry	120
	15.3 Exponentials (and Logarithms)	121

Bibliography 127

Index 129

List of Figures and Tables

2.1	A passenger on a train throws a ball to the right.	4
2.2	A lamp flashes on a moving train.	6
2.3	A lamp on a moving train as seen from the ground.	6
2.4	Time dilation by observing bouncing light.	7
2.5	The (ordinary) Pythagorean theorem for a light beam.	8
2.6	Length contraction by observing bouncing light.	9
3.1	Measuring distance in Euclidean geometry.	14
3.2	Defining (circular) trigonometric functions via unit circle.	15
3.3	A triangle with $\tan\theta = \frac{3}{4}$.	16
3.4	Projection.	16
3.5	A rotated coordinate system.	16
3.6	Width is coordinate-dependent.	17
3.7	The addition formula for slopes.	18
4.1	The graphs of $\cosh\beta$, $\sinh\beta$, and $\tanh\beta$.	20
4.2	Defining hyperbolic trigonometric functions via a hyperbola.	21
4.3	A hyperbolic triangle with $\tanh\beta = \frac{3}{5}$.	22
4.4	Hyperbolic projection.	23
5.1	Simple spacetime diagrams.	27
5.2	Lorentz transformations as hyperbolic rotations.	28
5.3	Causality.	30
5.4	Some hyperbolic right triangles.	33
5.5	More hyperbolic right triangles.	33

5.6 Hyperbolic projections of vectors I. 34
5.7 Hyperbolic projections of vectors II. 35

6.1 The Einstein addition formulas. 39
6.2 Length contraction as a hyperbolic projection. 40
6.3 Time dilation as a hyperbolic projection. 41
6.4 A three-dimensional spacetime diagram. 42
6.5 The spacetime Pythagorean theorem for a light beam. . . 43
6.6 The Doppler effect. 44

7.1 The spacetime diagram for a traveling muon. 47
7.2 The spacetime diagram for the rocket ship. 48
7.3 The spacetime diagram for the Lincoln and the VW. . . . 50
7.4 The spacetime diagram for the scientist's analysis. 50
7.5 Relative speeds in different reference frames 52
7.6 The angle made by the mast of a moving sailboat. 53
7.7 The angle made by a ball thrown from a moving sailboat. 54
7.8 The angle made by an antenna on a moving spaceship. . . 54
7.9 The angle made by the beam of a moving searchlight. . . 55
7.10 The spacetime diagram for Dr. X. 56
7.11 Cosmic rays. 58
7.12 Computing Doppler shift. 59

8.1 The pole and barn paradox. 62
8.2 The twin paradox 64

9.1 The geometry of proper time. 68
9.2 The geometry of 2-momentum. 72

10.1 The momentum diagram for two colliding lumps of clay. . 76
10.2 Momentum diagrams for the collision of identical particles. 78
10.3 Momentum diagram for the decay of a pion into photons. 79
10.4 Analyzing pion decay by using hyperbola geometry. 79
10.5 Spacetime diagram for the collision of identical particles. . 80
10.6 Spacetime diagram for the decay of a pion. 81

11.1 A wire with zero net charge density at rest. 84
11.2 A moving capacitor. 87

13.1 A ball thrown in a moving train, as seen from the train. . 100
13.2 A ball thrown in a moving train, as seen from the ground. 100

13.3 Tidal effects on falling objects. 101
13.4 How tides are caused by the earth falling toward the moon. 102
13.5 Classification of geometries. 103
13.6 The trajectory of a uniformly accelerating object. 106

14.1 The hyperboloid in three-dimensional Minkowski space. . 109
14.2 Measuring distance on the hyperboloid. 110
14.3 Stereographic projection of the hyperboloid. 111
14.4 Some hyperbolic lines in the Poincaré disk. 112
14.5 Constructing the Klein disk. 113
14.6 Some hyperbolic lines in the Klein disk. 114
14.7 The tractrix and the pseudosphere. 116
14.8 Embedding the pseudosphere in the hyperboloid. 116

15.1 The geometric definition of $\exp(\beta)$. 122
15.2 Using a shifted right triangle. 123
15.3 The geometry of the hyperbolic addition formulas. 124
15.4 The geometric verification that "exp" is exponential. . . . 125

Preface

The unification of space and time introduced by Einstein's special theory of relativity is one of the cornerstones of the modern scientific description of the universe. Yet the unification is counterintuitive because we perceive time very differently from space. Even in relativity, time is not just another dimension, it is one with different properties. Some authors have tried to "unify" the treatment of time and space, typically by replacing t by it, thus hiding some annoying minus signs. But these signs carry important information: our universe, as described by relativity, is *not* Euclidean.

This short book treats the geometry of hyperbolas as the key to understanding special relativity. This approach can be summarized succinctly as the replacement of the ubiquitous γ symbol of most standard treatments with the appropriate hyperbolic trigonometric functions. In most cases, this not only simplifies the appearance of the formulas, but emphasizes their geometric content in such a way as to make them almost obvious. Furthermore, many important relations, including but not limited to the famous relativistic addition formula for velocities, follow directly from the appropriate trigonometric addition formulas.

I am unaware of any other introductory book on special relativity that adopts this approach as fundamental. Many books point out the relationship between Lorentz transformations and hyperbolic rotations, but few actually make use of it. A pleasant exception is the original edition of Taylor and Wheeler's marvelous book [Taylor and Wheeler 63], but much of this material is not included in the second edition [Taylor and Wheeler 92].

The present book is not intended as a replacement for any of the excellent textbooks on special relativity. Rather, it is intended as an introduction to a particularly beautiful way of looking at special relativity, in hopes of encouraging students to see beyond the formulas to the deeper structure. Enough applications are included to get the basic idea, but these would probably need to be supplemented for a full course.

Much of the material presented can be understood by those familiar with the ordinary trigonometric functions, but occasional use is made of elementary differential calculus. In addition, the chapter on electricity and magnetism assumes the reader has seen Maxwell's equations and has at least a passing acquaintance with vector calculus. A prior course in calculus-based physics, up to and including electricity and magnetism, should provide the necessary background.

After a general introduction in Chapter 1, the basic physics of special relativity is described in Chapter 2. This is a quick, intuitive introduction to special relativity, and it sets the stage for the geometric treatment that follows. Chapter 3 summarizes some standard (and some not so standard) properties of ordinary two-dimensional Euclidean space, expressed in terms of the usual circular trigonometric functions; this geometry is referred to as *circle geometry*. This material has deliberately been arranged to closely parallel the treatment of two-dimensional Minkowski space in Chapter 4 in terms of hyperbolic trigonometric functions, which we call *hyperbola geometry*.[1] Special relativity is covered again from the geometric point of view in Chapters 5 and 6, followed by a discussion of some of the standard paradoxes in Chapter 8, applications to relativistic mechanics in Chapter 9, and the relativistic unification of electricity and magnetism in Chapter 11. Chapters 7, 10, and 12 present relevant problems and their solutions. Chapter 13 concludes our tour of relativity with a brief discussion of the further steps leading to Einstein's theory of general relativity. The final two chapters discuss applications of hyperbola geometry in mathematics, treating hyperbolic geometry in Chapter 14, and giving a geometric construction of the derivatives of trigonometric functions, as well as the exponential function and its derivative, in Chapter 15.

A companion website for this book is available at

http://physics.oregonstate.edu/coursewikis/GSR/bookinfo

This website contains further information about the course taught at Oregon State University, including links to syllabi, lecture outlines, and small group activities, and will be updated as needed with comments and corrections. Reader comments can also be entered at this website.

---

[1] Hyperbola geometry should not be confused with *hyperbolic geometry*, the curved geometry of the two-dimensional unit hyperboloid; see Chapter 14.

Acknowledgments

This book started out as class notes [Dray 03] for a course on *Reference Frames*, which in turn forms part of a major upper-division curriculum reform effort titled *Paradigms in Physics*, which was begun in the Department of Physics at Oregon State University in 1997. The class notes were subsequently published online in wiki format [Dray 10], and are now used as the primary text in that course; an abbreviated version was also published as a journal article [Dray 04].

I am grateful to all of the faculty involved in this effort, but especially to the leader of the project, my wife, Corinne Manogue, for support and encouragement at every stage. The *Paradigms in Physics* project has been supported in part by NSF grants DUE–965320, 0231194, 0618877, and 1023120, supplemented with funds from Oregon State University; my own participation was made possible thanks to the (sometimes reluctant) support of my department chair at the time, John Lee. I was fortunate in having excellent teaching assistants, Jason Janesky and Emily Townsend, when I first taught the course.

A course based on an early draft of this book was taught at Mount Holyoke College in 2002, giving me an opportunity to make further revisions; my stay at Mount Holyoke was partially supported by their Hutchcroft Fund. I am grateful to Greg Quenell for having carefully read the manuscript at that time and for suggesting improvements, and to Alex Brummer, who did the same more recently.

I thank David Griffiths, an old friend (and Corinne's former teacher), for permission to "borrow" homework problems from his standard textbook [Griffiths 99], as well as to include solutions using hyperbola geometry.

Last, but not least, I thank the many students who struggled to learn physics from a mathematician, enriching all of us.

Chapter 1

Introduction

1.1 Newton's Relativity

Our daily experience leads us to believe in Newton's laws. When you drop a ball, it falls straight down. When you throw a ball, it travels in a uniform (compass) direction—and falls down. We appear to be in a constant gravitational field, but apart from that no forces act on the ball. This isn't the full story, of course, because we are ignoring such things as air resistance and the spin of the ball. Nevertheless, Newton's laws seem to give a pretty good description of what we observe, and so we base our intuitive understanding of physics on them.

But Newton's laws are wrong.

Yes, gravity is more complicated than this simple picture. The earth's gravitational field isn't really constant. Furthermore, other nearby objects act gravitationally on us, notably the moon. As discussed in the final chapter of this book, this action causes tides.

A bigger problem is that the earth is round. Anyone who flies from San Francisco to New York is aware that due east is *not* a straight line, defined in this case as the shortest distance between two points. In fact, if you travel in a straight line that (initially) points due east from my home in Oregon, you will eventually pass to the south of the southern tip of Africa![1]

So east is not east.

But the real problem is that the earth is rotating. Try playing catch on a merry-go-round! Balls certainly don't seem to travel in a straight line! Newton's laws don't work here, and, strictly speaking, they don't work

---

[1] You can check this by stretching a string on a globe so that it goes all the way around, is as tight as you can make it, and goes through Oregon in an east/west direction.

on (that is, in the reference frame of) the earth's surface. The motion of a Foucault pendulum can be thought of as a Coriolis effect caused by an external *pseudoforce*. Also, a plumb bob doesn't actually point toward the center of the earth!

So down is not down.

1.2 Einstein's Relativity

All of the above problems come from the fact that, even without worrying about gravity, the earth's surface is *not* an inertial frame. An inertial frame is, roughly speaking, one in which Newton's laws do hold. Playing catch on a train is little different from on the ground—at least in principle, as long as the train is not speeding up or slowing down. Furthermore, an observer on the ground would see nothing out of the ordinary, it merely being necessary to combine the train's velocity with that of the ball.

However, shining a flashlight on a moving train—especially the description of this from the ground—turns out to be another story, which we study in more detail in Chapters 2 and 6. Light doesn't behave the way balls do, and this difference forces a profound change in our description of the world around us. As we will see, this forces moving objects to change in unexpected ways: their clocks slow down, they change size, and, in a certain sense, they get heavier.

So time is not time.

Of course, these effects are not very noticeable in our daily lives any more than Coriolis forces affect a game of catch. But some modern conveniences, notably global positioning technology, are affected by relativistic corrections.

The bottom line is that the reality is quite different from what our intuition says it ought to be. The world is neither Euclidean nor Newtonian. Special relativity isn't just some bizarre theory; it is a correct description of nature (ignoring gravity). It is also a beautiful theory, as I hope you will agree. Let's begin.

CHAPTER 2

THE PHYSICS OF SPECIAL RELATIVITY

In which it is shown that time is not the same for all observers.

2.1 OBSERVERS AND MEASUREMENT

This chapter provides a very quick introduction to the physics of special relativity, intended as a review for those who have seen it before, and as an overview for those who have not. It is not a prerequisite for the presentations in subsequent chapters.

Special relativity involves comparing what different observers see. But we need to be careful about what these words mean.

A *reference frame* is a way of labeling each event with its location in space and the time at which it occurs. Making a measurement corresponds to recording these labels for a particular event. When we say that an *observer* "sees" something, what we really mean is that a particular event is recorded in the reference frame associated with the observer. This has nothing to do with actually *seeing* anything—a much more complicated process that would involve keeping track of the light reflected into the observer's eyes. Rather, an "observer" is really an entire army of observers who record any interesting events, and an "observation" consists of reconstructing from their journals what took place.

2.2 THE POSTULATES OF SPECIAL RELATIVITY

Postulate I is the most fundamental postulate of relativity.

POSTULATE I. *The laws of physics apply in all inertial reference frames.*

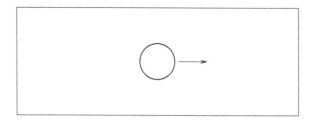

FIGURE 2.1. A passenger on a train throws a ball to the right. On an ideal train, it makes no difference whether the train is moving.

The first ingredient here is a class of preferred reference frames. Simply put, an inertial reference frame is one without external forces. More precisely, an inertial frame is one in which an object initially at rest will remain at rest. Because of gravity, inertial frames must be in free fall; consider a spaceship with its drive turned off or a falling elevator. Gravity causes additional complications, such as tidal effects, which force such freely falling frames to be small (compared to, say, the earth); we revisit this concept in Chapter 13. But special relativity describes a world without gravity, so in practice we describe inertial frames in terms of relative motion at constant velocity, typically in the form of an idealized train.

Applied to mechanics, Postulate I is the principle of *Galilean relativity*. For instance, consider a ball thrown to the right with speed u. Ignoring forces such as gravity and air friction, the ball keeps moving at the same speed forever, since there are no forces acting on it. Try the same thing on a train, which is itself moving to the right with speed v. Then Galilean relativity leads to the same conclusion: as seen from the train, the ball moves to the right with speed u forever. An observer on the ground, of course, sees the ball move with speed $u + v$; Galilean relativity insists only that both observers observe the same physics, namely, the lack of acceleration due to the absence of any forces, but not necessarily the same speed. This situation is shown in Figure 2.1.

Einstein generalized Postulate I by applying it not just to mechanics but also to electrodynamics. However, Maxwell's equations, presented in roughly their present form in the early 1860s, make explicit reference to the speed of light. In MKS units, Gauss's law (see Equation (11.72)) involves the permittivity constant ϵ_0, and Ampère's Law (see Equation (11.75)) involves the permeability constant μ_0; both of these can be measured experimentally. But Maxwell's equations predict electromagnetic waves—

2.2. THE POSTULATES OF SPECIAL RELATIVITY

including light—with a speed (in vacuum) of

$$c = \frac{1}{\sqrt{\epsilon_0 \mu_0}}. \tag{2.1}$$

Thus, from some relatively simple experimental data, Maxwell's equations *predict* that the speed of light in vacuum is

$$c = 3 \times 10^8 \text{ m/s}. \tag{2.2}$$

The famous Michelson-Morley experiment in 1887, at what is now Case Western Reserve University, set out to show that this speed is relative to the *ether*, so we should be able to measure our own motion relative to the ether by measuring direction-dependent variations in c. Instead, the experiment showed that there were no such variations; Einstein argued that there is therefore no ether![1] Postulate I together with Maxwell's equations therefore lead to

POSTULATE II. *The speed of light is the same for all inertial observers.*

The theory of special relativity follows from Postulates I and II, which were introduced by Einstein in 1905 [Einstein 05].[2] As we show next, an immediate consequence of these postulates is that two inertial observers disagree about whether two events are simultaneous.

Thought experiments have been around since at least the time of Euclid. However, the German phrase *Gedankenexperiment* was not coined until the early 1800s, and was first translated into English in the late 1800s, not long before Einstein's discovery of special relativity. Einstein himself attributed his breakthrough to thought experiments, including a childhood attempt to imagine what one would see if one could ride along with a beam of light. However, it remains unclear exactly what path he took to reach his conclusions about the observer-dependence of simultaneity. One possible path [Norton 06] is the following thought experiment.

Consider a train at rest with a lamp in the middle, as shown in Figure 2.2. After the light is turned on, light reaches both ends of the train at the same time, having traveled in both directions at constant speed c. Now try the same experiment on a moving train. This is still an inertial frame,

---

[1] Of historical interest is the fact that this was not the interpretation given by Michelson and Morley, who instead argued in favor of a now-discredited theory involving the dragging of the ether by the earth. Nonetheless, in 1907 Albert Abraham Michelson became the first American to win the Nobel Prize in Physics, in part due to this work.

[2] Albert Einstein won the Nobel Prize in Physics in 1921, but not for his work on special relativity.

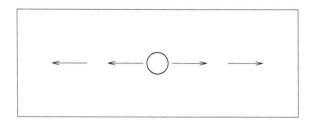

FIGURE 2.2. The same situation as for Figure 2.1, with the ball replaced by a lamp in the exact middle of the train. The light from the lamp reaches both ends of the train at the same time, whether the train is moving or at rest.

and so, just as with the ball in the previous example, one obtains the same result, namely, that the light reaches both ends of the train at the same time *as seen by an observer on the train*. However, the second postulate leads to a very different result for an observer on the ground. According to this postulate, the light travels at speed c as seen from the ground, *not* the expected $c \pm v$. But, as seen from the ground, the ends of the train also move while the light is getting from the middle of the train to the ends. The rear wall "catches up" with the approaching light beam, whereas the front wall "runs away." As shown in Figure 2.3, the net result is that the ground-based observer sees the light reach the rear of the train before it reaches the front; these two observers therefore disagree about whether the light reaches both ends of the train simultaneously.

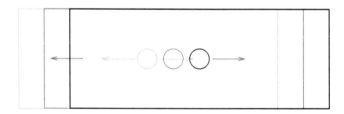

FIGURE 2.3. An observer on the ground sees the train go past with its lamp, as in Figure 2.2. But since the ground-based observer must also see the light travel with speed c, and since the ends of the train are moving while the light is traveling, this observer concludes that the light reaches the rear of the train before it reaches the front.

2.3. TIME DILATION AND LENGTH CONTRACTION

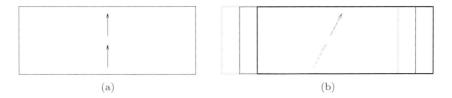

FIGURE 2.4. A beam of light bounces up and down between mirrors on the floor and ceiling of a moving train. The time between bounces can be used as a unit of time, but what is seen by (a) a moving observer on the train and (b) a stationary observer on the ground is quite different.

2.3 TIME DILATION AND LENGTH CONTRACTION

We have seen that the postulates of relativity force the surprising conclusion that time is observer-dependent. We now examine this phenomenon in more detail.

Consider again a train, of height h, with a beam of light bouncing up and down between mirrors on the floor and ceiling, as shown in Figure 2.4(a). The time between bounces can be interpreted as the "ticks" of a clock, and this interval, *as measured on the train*, is independent of whether the train is moving. However, as shown in Figure 2.4(b), a stationary observer on the ground sees something quite different.

From the ground, the light appears to move diagonally, and hence travels a longer path than the vertical path seen on the train. But since the light must move at the same speed for both observers, each "tick" takes longer according to the ground-based observer than it does for the observer on the train. Thus, the observer at rest sees the "clock" of the moving observer run slow.

Work through each step of this argument carefully; the key assumption is Postulate II—namely, that the light must travel at the same speed for both observers. But this is not the behavior we expect from our daily experience.

To compute how the times are related, we must first introduce some notation. Let t denote time as measured on the ground, and t' denote time as measured on the train; we similarly use x and x' to measure length. One tick of the clock as seen on the train takes time $\Delta t'$, where the distance traveled is

$$h = c\,\Delta t'. \tag{2.3}$$

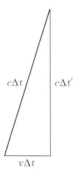

FIGURE 2.5. The (ordinary) Pythagorean theorem for a bouncing light beam on a moving train.

Suppose the same tick takes time Δt as seen from the ground. In this time, the light travels a distance $c\,\Delta t$, which is the hypotenuse of a right triangle with legs h and $v\,\Delta t$. As shown in Figure 2.5, the Pythagorean theorem now leads to

$$(c\,\Delta t)^2 = (v\,\Delta t)^2 + h^2, \tag{2.4}$$

which can be solved for h. Comparing this result with that of Equation (2.3), we obtain

$$\Delta t = \frac{1}{\sqrt{1 - \frac{v^2}{c^2}}}\,\Delta t', \tag{2.5}$$

which indeed shows that the moving clock runs slower than the stationary one ($\Delta t > \Delta t'$), at least for speeds $v < c$. The factor relating these times shows up so often that we give it a special symbol,

$$\gamma = \frac{1}{\sqrt{1 - \frac{v^2}{c^2}}}. \tag{2.6}$$

This effect, called *time dilation*, has important consequences for objects traveling at a significant fraction of the speed of light, but has virtually no effect for objects at everyday speeds ($v \ll c$). This is the effect that allows cosmic rays to reach the earth—the particles' lifetimes, as measured by their own clocks, is many orders of magnitude shorter than the time we observe them traveling through the atmosphere.

Time is not the only thing on which observers do not agree. Consider now a beam of light bouncing horizontally between the front and back of

2.3. Time Dilation and Length Contraction

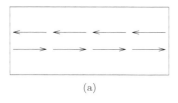 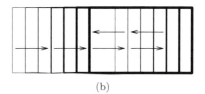

(a)   (b)

FIGURE 2.6. A beam of light bounces back and forth between mirrors at the front and back of a moving train. The time of the roundtrip journey can be used to measure the length of the train, but (a) a moving observer on the train and (b) a stationary observer on the ground obtain different results.

the train, as shown in Figure 2.6(a). As seen on the train, if it takes time $\Delta t'$ to make a round trip, and the length of the train is $\Delta x'$, then we must have

$$c\,\Delta t' = 2\,\Delta x'. \tag{2.7}$$

What does the observer on the ground see? Don't forget that the train is moving, so that, as shown in Figure 2.6(b), the distance traveled in one direction is different from that in the other. More precisely, light starting from the back of the train must "chase" the front; if it takes time Δt_1 to catch up, then the distance traveled is $\Delta x + v\,\Delta t_1$, the sum of the length of the train (as seen from the ground) and the distance the front of the train traveled while the light was under way. Similarly, if the time taken on the return journey is Δt_2, then the distance traveled is $\Delta x - v\,\Delta t_2$. Thus,

$$c\,\Delta t_1 = \Delta x + v\,\Delta t_1 \tag{2.8}$$

and

$$c\,\Delta t_2 = \Delta x - v\,\Delta t_2, \tag{2.9}$$

or, equivalently,

$$\Delta t_1 = \frac{\Delta x}{c - v} \tag{2.10}$$

and

$$\Delta t_2 = \frac{\Delta x}{c + v}. \tag{2.11}$$

Combining these results leads to

$$c\,\Delta t = c\,\Delta t_1 + c\,\Delta t_2$$
$$= \frac{2\,\Delta x}{1 - \frac{v^2}{c^2}}, \qquad (2.12)$$

and combining Equations (2.12), (2.5), and (2.7) then leads to

$$\Delta x = \sqrt{1 - \frac{v^2}{c^2}}\,\Delta x' = \frac{1}{\gamma}\,\Delta x'. \qquad (2.13)$$

Thus, a moving object appears to be shorter in the direction of motion than it would be at rest; this effect is known as *length contraction*.

2.4 LORENTZ TRANSFORMATIONS

Suppose observer O' is moving to the right with speed v. If x denotes the distance of an object from a stationary observer O,[3] then the distance between the object and O' as measured by O will be $x - vt$. But using the formula for length contraction (2.13), namely, $\Delta x' = \gamma\,\Delta x$, we see that

$$x' = \gamma\,(x - vt). \qquad (2.14)$$

By Postulate I, the framework used by each observer to describe the other must be the same. In particular, if we interchange the roles of O and O', nothing else should change—except for the fact that the relative velocity (of O with respect to O') is now $-v$ instead of v. By symmetry, we therefore have immediately that

$$x = \gamma\,(x' + vt'), \qquad (2.15)$$

where we have been careful not to assume that $t' = t$. In fact, combining Equations (2.14) and (2.15) quickly yields

$$t' = \gamma\left(t - \frac{v}{c^2}x\right) \qquad (2.16)$$

and a similar expression for t in terms of x' and t'.

Equations (2.14) and (2.16) together give the *Lorentz transformation* from the frame of the observer a rest (O) to the frame of the moving observer (O').[4] By symmetry, we can invert these expressions simply by replacing v with $-v$.

---

[3]This distance need not be constant, but could be a function of time.
[4]Lorentz transformations are named after Hendrik Lorentz, a Dutch physicist who won the 1902 Nobel Prize in Physics.

2.5 Addition of Velocities

Suppose that, as seen from O, O' is moving to the right with speed v and that an object is moving to the right with speed u. According to Galileo, we would simply add velocities to determine the velocity of the object as seen from O':

$$u = u' + v. \tag{2.17}$$

Equation (2.17) can be derived by differentiating the Galilean transformation

$$x = x' + vt \tag{2.18}$$

with respect to t, thus obtaining

$$\frac{dx}{dt} = \frac{dx'}{dt} + v. \tag{2.19}$$

To derive the relativistic formula for the addition of velocities, we proceed similarly. However, since *both* x and t transform, it is useful to use the differential form of the Lorentz transformations, namely,

$$dx = d\left(\gamma\left(x' + vt'\right)\right) = \gamma\left(dx' + v\,dt'\right), \tag{2.20}$$

$$dt = d\left(\gamma\left(t' + \frac{v}{c^2}x'\right)\right) = \gamma\left(dt' + \frac{v}{c^2}dx'\right). \tag{2.21}$$

Dividing expression (2.20) by (2.21) leads to

$$\frac{dx}{dt} = \frac{\frac{dx'}{dt'} + v}{1 + \frac{v}{c^2}\frac{dx'}{dt'}}, \tag{2.22}$$

or, equivalently,

$$u = \frac{u' + v}{1 + \frac{u'v}{c^2}}. \tag{2.23}$$

Equation (2.23) is known as the *Einstein addition formula*, and shows that the addition of relativistic velocities does not obey our intuitive notion of how velocities should add. It is worth considering special cases of this formula, such as when both u' and v are small compared to c, or when one of them is equal to c. We provide a geometric explanation for these apparently peculiar properties in Chapter 6.

2.6 THE INTERVAL

Direct computation using the Lorentz transformations shows that

$$\begin{aligned}x'^2 - c^2 t'^2 &= \gamma^2 (x - vt)^2 - \gamma^2 \left(ct - \frac{v}{c}x\right)^2 \\ &= x^2 - c^2 t^2,\end{aligned} \quad (2.24)$$

so that the quantity $x^2 - c^2 t^2$, known as the *interval*, does not depend on the observer who computes it. We explore this concept further in later chapters.

This concludes our brief introduction to the underlying physics. In subsequent chapters we will recover these results from a more geometric point of view.

CHAPTER 3

CIRCLE GEOMETRY

In which some standard properties of two-dimensional Euclidean geometry are reviewed, and some more subtle properties are pointed out.

3.1 THE GEOMETRY OF TRIGONOMETRY

We begin with a review of the geometry of trigonometry. Before reading further, you are encouraged to reflect briefly on the most important features of trigonometry. Make a list! As you then read through this chapter, compare your list to the properties discussed here. At the end of the chapter, ask yourself whether your list has changed. (You may wish to repeat this exercise as you read Chapter 4.)

3.2 DISTANCE

The key concept in Euclidean geometry is the *distance function*, which measures the distance between two points. In two dimensions, the squared distance between the point B with coordinates (x, y) and the origin is given by

$$r^2 = x^2 + y^2, \tag{3.1}$$

which is, of course, just the Pythagorean theorem; see Figure 3.1.

3.3 CIRCLE TRIGONOMETRY

It is natural to study the set of points that are a constant distance from a given point, which of course form a *circle*.

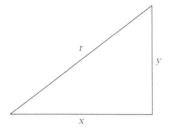

FIGURE 3.1. Measuring distance in Euclidean geometry by using the Pythagorean theorem, which states that $r^2 = x^2 + y^2$.

Consider a point P on the circle of radius r shown in Figure 3.2. The angle θ between the line from the origin to P and the (positive) x-axis is *defined* as the ratio of the length the arc of a circle between P and the point $(r, 0)$ to the radius r.[1] Denoting the coordinates of P by (x, y), the basic (circular) trigonometric functions are then *defined* by

$$\cos\theta = \frac{x}{r}, \qquad (3.2)$$

$$\sin\theta = \frac{y}{r}, \qquad (3.3)$$

$$\tan\theta = \frac{\sin\theta}{\cos\theta}, \qquad (3.4)$$

and the fundamental identity

$$\cos^2\theta + \sin^2\theta = 1 \qquad (3.5)$$

follows from the definition of a circle. We then have the well-known addition formulas[2]

$$\sin(\theta + \phi) = \sin\theta\cos\phi + \cos\theta\sin\phi, \qquad (3.6)$$

$$\cos(\theta + \phi) = \cos\theta\cos\phi - \sin\theta\sin\phi, \qquad (3.7)$$

$$\tan(\theta + \phi) = \frac{\tan\theta + \tan\phi}{1 - \tan\theta\tan\phi}, \qquad (3.8)$$

---

[1] This definition is commonly stated in terms of the *unit* circle, a practice we avoid in order to more easily distinguish dimensionless quantities such as angles from dimensionful quantities such as lengths.

[2] A particularly elegant proof of Equation (3.7) is to compute the dot product of two unit vectors both algebraically and geometrically, and compare the results. Equations (3.6) and (3.8) follow easily from (3.7); see, for example, [Dray and Manogue 06].

3.4. TRIANGLE TRIGONOMETRY

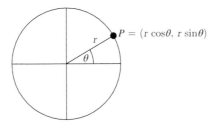

FIGURE 3.2. Defining the (circular) trigonometric functions via the unit circle.

as well as the derivative formulas[3]

$$\frac{d}{d\theta} \sin \theta = \cos \theta, \qquad (3.9)$$

$$\frac{d}{d\theta} \cos \theta = -\sin \theta. \qquad (3.10)$$

3.4 TRIANGLE TRIGONOMETRY

An important class of trigonometric problems involves determining, say, $\cos \theta$ if $\tan \theta$ is known. One can, of course, do this algebraically by using the identity

$$\cos^2 \theta = \frac{1}{1 + \tan^2 \theta}. \qquad (3.11)$$

It is often easier to do this geometrically, as illustrated by the following example.

Suppose you know $\tan \theta = \frac{3}{4}$, and you wish to determine $\cos \theta$. Draw *any* triangle containing an angle whose tangent is $\frac{3}{4}$. In this case, the obvious choice would be the triangle shown in Figure 3.3, with sides of 3 and 4. What then is $\cos \theta$? The hypotenuse clearly has length 5, so $\cos \theta = \frac{4}{5}$.

Trigonometry is not merely about ratios of sides, it is also about projections. Another common use of triangle trigonometry is to determine the sides of a triangle given the hypotenuse r and one angle θ. The answer, of course, is that the sides are $r \cos \theta$ and $r \sin \theta$, as shown in Figure 3.4.

---

[3] An alternate derivation of these derivative formulas is given in Section 15.1.

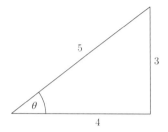 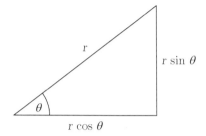

FIGURE 3.3. A triangle with $\tan\theta = \frac{3}{4}$.

FIGURE 3.4. A triangle in which the hypotenuse and one angle are known.

3.5 ROTATIONS

Now consider a new set of coordinates, call them (x', y'), based on axes rotated counterclockwise through an angle θ from the original ones, as shown in Figure 3.5. For instance, the y-axis could point toward true north, whereas the y'-axis might point toward magnetic north. In the primed coordinate system, point B has coordinates $(r, 0)$, and point A has coordinates $(0, r)$, whereas the unprimed coordinates of B and A are, respectively, $(r\cos\theta, r\sin\theta)$ and $(-r\sin\theta, r\cos\theta)$. A little work shows that the coordinate systems are related by a rotation matrix

$$\begin{pmatrix} x \\ y \end{pmatrix} = \begin{pmatrix} \cos\theta & \sin\theta \\ -\sin\theta & \cos\theta \end{pmatrix} \begin{pmatrix} x' \\ y' \end{pmatrix} \qquad (3.12)$$

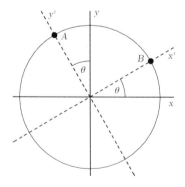

FIGURE 3.5. A rotated coordinate system.

(where we are assuming that all coordinates are measured in the same units). Furthermore, both coordinate systems lead to the same result for the squared distance, since

$$x^2 + y^2 = x'^2 + y'^2. \tag{3.13}$$

3.6 PROJECTIONS

Consider the rectangular object of width 1 meter shown in Figure 3.6(a), which has been rotated so it is parallel to the *primed* axes in Figure 3.5. How wide is it? As worded, this question is poorly posed. If width means "extent in the x' direction," then the answer is 1 meter. If, however, width means "extent in the x direction," then the answer is obtained by measuring the *horizontal* distance between the sides of the rectangle, which results in a value *larger* than 1. (The exact value is easily seen to be $1/\cos\theta$.)

Repeat this exercise in the opposite direction. Take the same rectangular object, but orient it parallel to the *unprimed* axes, as shown in Figure 3.6(b). How wide is it? Clearly the "unprimed" width is 1 meter, and the "primed" width is larger (again given by $1/\cos\theta$).

In one of the cases above, the "primed" width is smaller, yet in the other the "unprimed" width is smaller. What is happening here? If you turn your suitcase at an angle, it is *harder* to fit under your seat! It has, in effect, become "longer." But which orientation is best depends, of course, on the orientation of the seat.

Remember this discussion when we address the corresponding questions in relativity in subsequent chapters.

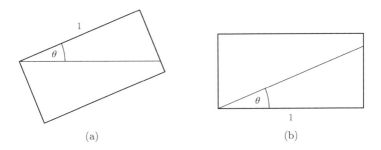

FIGURE 3.6. Width is coordinate-dependent. The rectangle is oriented (a) parallel to the primed axes of Figure 3.5 and (b) parallel to the unprimed axes.

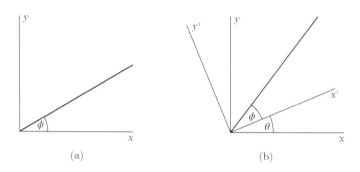

FIGURE 3.7. The addition formula for slopes. The line makes an angle ϕ with (a) the x-axis and (b) the x'-axis.

3.7 ADDITION FORMULAS

Consider the line through the origin that makes an angle ϕ with the positive x-axis, as shown in Figure 3.7(a). What is its slope? The equation of the line is

$$y = x \tan\phi, \tag{3.14}$$

so that the slope is $\tan\phi$, at least in "unprimed" coordinates. Consider now the line through the origin shown in Figure 3.7(b), which makes an angle ϕ with the positive x'-axis. What is its slope? In "primed" coordinates, the equation of the line is just

$$y' = x' \tan\phi, \tag{3.15}$$

so that, in these coordinates, the slope is again $\tan\phi$. But what about in "unprimed" coordinates? The x' axis itself makes an angle θ with the x-axis. It is tempting to simply add these slopes, obtaining $\tan\phi + \tan\theta$, but this is not correct. Slopes don't add, angles do. The correct answer is that

$$y = x \tan(\theta + \phi), \tag{3.16}$$

so that the slope is given by Equation (3.8).

Remember this discussion when we discuss the Einstein addition formula.

➤ Chapter 4 ≺

Hyperbola Geometry

In which a two-dimensional non-Euclidean geometry is constructed, which will turn out to be identical with that of special relativity.

4.1 Hyperbolic Trigonometry

The hyperbolic trigonometric functions are usually defined by using the formulas

$$\cosh \beta = \frac{e^\beta + e^{-\beta}}{2}, \tag{4.1}$$

$$\sinh \beta = \frac{e^\beta - e^{-\beta}}{2}, \tag{4.2}$$

$$\tanh \beta = \frac{\sinh \beta}{\cosh \beta}, \tag{4.3}$$

and so on. We give an alternative definition in the next section. The graphs of these hyperbolic functions are shown in Figure 4.1. It is straightforward to verify from these definitions that

$$\cosh^2 \beta - \sinh^2 \beta = 1, \tag{4.4}$$

$$\sinh(\alpha + \beta) = \sinh \alpha \cosh \beta + \cosh \alpha \sinh \beta, \tag{4.5}$$

$$\cosh(\alpha + \beta) = \cosh \alpha \cosh \beta + \sinh \alpha \sinh \beta, \tag{4.6}$$

$$\tanh(\alpha + \beta) = \frac{\tanh \alpha + \tanh \beta}{1 + \tanh \alpha \tanh \beta}, \tag{4.7}$$

$$\frac{d}{d\beta} \sinh \beta = \cosh \beta, \tag{4.8}$$

$$\frac{d}{d\beta} \cosh \beta = \sinh \beta. \tag{4.9}$$

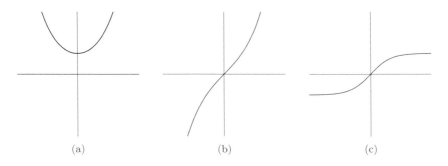

FIGURE 4.1. The graphs of (a) $\cosh\beta$, (b) $\sinh\beta$, and (c) $\tanh\beta$.

A geometric derivation of these properties is given in Chapter 15.

The properties of the hyperbolic trigonometric functions look very much like their ordinary trigonometric counterparts (except for signs). This similarity derives from the identities

$$\cosh\beta \equiv \cos(i\beta), \qquad (4.10)$$
$$\sinh\beta \equiv -i\sin(i\beta). \qquad (4.11)$$

4.2 DISTANCE

We saw in Chapter 3 that Euclidean trigonometry is based on *circles*, sets of points that are a constant distance from the origin. Hyperbola geometry is obtained simply by using a different distance function. Measure the "squared distance" of the point B with coordinates (x, y) from the origin by using the definition

$$\rho^2 = x^2 - y^2. \qquad (4.12)$$

This distance function is the key idea in hyperbola geometry.

4.3 HYPERBOLA TRIGONOMETRY

In hyperbola geometry, "circles" of constant distance from the origin become hyperbolas with $\rho = $ constant. We further restrict ourselves to the branch with $x > 0$. If B is a point on this hyperbola, then we can *define* the hyperbolic angle β between the line from the origin to B and the positive

4.3. Hyperbola Trigonometry

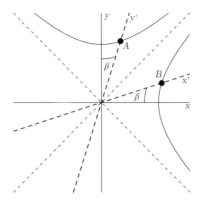

FIGURE 4.2. Defining the hyperbolic trigonometric functions via a hyperbola. Point A has coordinates $(\rho \sinh\beta, \rho \cosh\beta)$, and point B has coordinates $(\rho \cosh\beta, \rho \sinh\beta)$, where ρ is the *hyperbolic* radius.

x-axis to be the ratio of the *Lorentzian* length[1] of the arc of the hyperbola between B and the point $(\rho, 0)$ to the "radius" ρ. We could then *define* the hyperbolic trigonometric functions in terms of the coordinates (x, y) of B; that is,

$$\cosh\beta = \frac{x}{\rho}, \qquad (4.13)$$

$$\sinh\beta = \frac{y}{\rho}. \qquad (4.14)$$

A little work shows that this definition is equivalent to the one given by Equations (4.1) and (4.2).[2] This construction is shown in Figure 4.2, which also shows another hyperbola, given by $x^2 - y^2 = -\rho^2$. By symmetry, point A on this hyperbola has coordinates $(x, y) = (\rho \sinh\beta, \rho \cosh\beta)$. The importance of this hyperbola will become clear in Chapter 6.

---

[1] No, we haven't defined this. In Euclidean geometry, the length of a curve is obtained by integrating ds along the curve, where $ds^2 = dx^2 + dy^2$. In a similar way, the Lorentzian length is obtained by integrating $d\sigma$, where $d\sigma^2 = |dx^2 - dy^2|$.

[2] Use $x^2 - y^2 = \rho^2$ to compute

$$\rho^2 \, d\beta^2 \equiv d\sigma^2 = dy^2 - dx^2 = \frac{\rho^2 \, dx^2}{x^2 - \rho^2} = \frac{\rho^2 \, dy^2}{y^2 + \rho^2},$$

then take the square root of either expression and integrate. (The integrals are hard.) Finally, solving for x or y in terms of β yields Equation (4.1) or (4.2), respectively, and the other equation then follows immediately from (4.4).

Many of the features of the graphs shown in Figure 4.1 follow immediately from this definition of the hyperbolic trigonometric functions in terms of coordinates along the original hyperbola. Since the minimum value of x on this hyperbola is ρ, we must have $\cosh\beta \geq 1$. As β approaches $\pm\infty$, x approaches ∞ and y approaches $\pm\infty$, which agrees with the asymptotic behavior of the graphs of $\cosh\beta$ and $\sinh\beta$, respectively. Finally, since the hyperbola has asymptotes $y = \pm x$, we see that $|\tanh\beta| < 1$, and that $\tanh\beta$ must approach ± 1 as β approaches $\pm\infty$.

So how do we measure the distance between two points? The squared distance defined in Equation (4.12) can be positive, negative, or zero. We thus adopt the following convention: *Take the square root of the absolute value of the squared distance.* As shown in Chapter 5, it is important to remember whether the squared distance is positive or negative, but this corresponds directly to whether the distance is "mostly horizontal" or "mostly vertical."

4.4 TRIANGLE TRIGONOMETRY

We now recast ordinary triangle trigonometry into hyperbola geometry. Suppose you know $\tanh\beta = \frac{3}{5}$, and you wish to determine $\cosh\beta$. One can, of course, do this algebraically by using the identity

$$\cosh^2\beta = \frac{1}{1-\tanh^2\beta}. \tag{4.15}$$

But it is easier to draw *any* triangle containing an angle whose hyperbolic tangent is $\frac{3}{5}$. In this case, the obvious choice would be the triangle shown in Figure 4.3, with sides of 3 and 5.

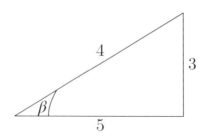

FIGURE 4.3. A hyperbolic triangle with $\tanh\beta = \frac{3}{5}$.

4.5. ROTATIONS

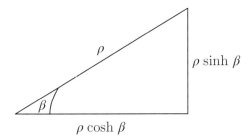

FIGURE 4.4. A hyperbolic triangle in which the hypotenuse and one angle are known.

What is $\cosh\beta$? We first need to work out the length ρ of the hypotenuse. The (hyperbolic) Pythagorean theorem tells us that

$$5^2 - 3^2 = \rho^2, \tag{4.16}$$

so ρ is clearly 4. Take a good look at the 3–4–5 triangle of hyperbola geometry shown in Figure 4.3. Now that we know all the sides of the triangle, it is easy to see that $\cosh\beta = \frac{5}{4}$.

Trigonometry is not merely about the ratios of the sides of triangles; it is also about projections. Another common use of triangle trigonometry is to determine the sides of a triangle given the hypotenuse ρ and one angle β. The answer, of course, is that the sides are $\rho\cosh\beta$ and $\rho\sinh\beta$, as shown in Figure 4.4.

4.5 ROTATIONS

By analogy with the Euclidean case, we *define* a hyperbolic rotation through the relations

$$\begin{pmatrix} x \\ y \end{pmatrix} = \begin{pmatrix} \cosh\beta & \sinh\beta \\ \sinh\beta & \cosh\beta \end{pmatrix} \begin{pmatrix} x' \\ y' \end{pmatrix}. \tag{4.17}$$

This corresponds to "rotating" both the x- and y-axes into the first quadrant, as shown in Figure 4.2. While this may seem peculiar, it is easily verified that the "distance" is invariant, that is,

$$x^2 - y^2 \equiv x'^2 - y'^2, \tag{4.18}$$

which follows immediately from the hyperbolic trigonometric identity (4.4).

4.6 PROJECTIONS

We can ask the same question as we did for Euclidean geometry: consider a rectangle of width 1 whose sides are parallel to the unprimed axes; how wide is it when measured in the primed coordinates? It turns out that the width of the box in the primed coordinate system is *less than* 1. This is length contraction, to which we will return in Chapter 6, along with time dilation.

4.7 ADDITION FORMULAS

What is the slope of the line from the origin to point A in Figure 4.2? The equation of this line, the y'-axis, is

$$x = y \tanh \beta. \tag{4.19}$$

Consider now a line with equation

$$x' = y' \tanh \alpha. \tag{4.20}$$

What is its (unprimed) slope? Again, slopes don't add, but hyperbolic angles do; therefore,

$$x = y \tanh(\alpha + \beta), \tag{4.21}$$

which can be expressed in terms of the slopes $\tanh \alpha$ and $\tanh \beta$ by using Equation (4.7). As discussed in more detail in Chapter 6, this is the Einstein addition formula.

➢ CHAPTER 5 ≺

THE GEOMETRY OF SPECIAL RELATIVITY

In which it is shown that special relativity is just hyperbola geometry.

5.1 THE SURVEYORS

A brilliant aid in understanding special relativity is the *surveyors' parable* introduced by Taylor and Wheeler [Taylor and Wheeler 63, Taylor and Wheeler 92]. Suppose a town has daytime surveyors, who determine north and east with a compass, and nighttime surveyors, who use the North Star. These notions differ, of course, since magnetic north is not the direction to the North Pole. Suppose, further, that both groups measure north/south distances in miles and east/west distances in meters, with both being measured from the town center. How does one go about comparing the measurements of the two groups?

With our knowledge of Euclidean geometry, we see how to do this: convert miles to meters (or vice versa). Distances computed with the Pythagorean theorem do not depend on which group does the surveying. Finally, it is easily seen that "daytime" coordinates can be obtained from "nighttime" coordinates by a simple rotation. The geometry of this situation is therefore described by Figure 3.5, in which the x and y directions correspond to geographic east and north, respectively, and the x' and y' directions correspond to *magnetic* east and north, respectively. If the surveyors measure x and x' in meters, and y and y' in miles, and if they do not understand how to convert between the two, communication between the two groups will not be easy.

The moral of this parable is therefore:

1. *The same units should be used for all distances.*

2. *The (squared) distance is invariant.*

3. *Different frames are related by rotations.*

Applying this lesson to special relativity, we should measure both time and space in the same units. How do we measure distance in seconds? That's easy: simply multiply by c. Thus, since $c = 3 \times 10^8$ m/s, 1 second of distance is just 3×10^8 m.[1] This has the effect of setting $c = 1$, since the number of seconds (of distance) traveled by light in 1 second (of time) is precisely 1.

Of course, it is also possible to measure time in meters: simply divide by c. Thus, 1 meter of time is the time it takes for light (in vacuum) to travel 1 meter. Again, this has the effect of setting $c = 1$.

5.2 SPACETIME DIAGRAMS

A *spacetime diagram* is simply a diagram showing both where things are and when they were there. For example, Figure 5.1(a) represents somebody sitting still (but getting older!), and Figure 5.1(b) represents somebody moving to the right at constant speed. Such trajectories are called *worldlines*.[2] In relativity, the convention is to have the t-axis be vertical, and the x-axis be horizontal.

The surveyors' parable tells us that we should measure space and time in the same units, not one in meters and the other in seconds, but both in either meters or seconds. We choose meters; this amounts to using ct to measure time rather than t. In either case, light beams play a special role in spacetime diagrams because they are drawn at 45°.

One fundamental geometric difference between circle trigonometry and hyperbola trigonometry is the presence of asymptotes in the hyperbolic case. These asymptotes have physical significance. Recall that inverse slopes in a spacetime diagram correspond to speeds; the asymptotes correspond to the existence of a "special" speed, namely, the speed of light.

---

[1] A similar unit of distance is the *lightyear*, namely, the distance traveled by light in 1 year, which would here be called simply a *year* of distance.

[2] One early French railroad timetable actually shows the worldlines of each train.

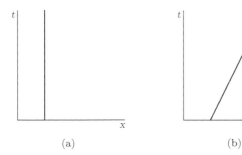

FIGURE 5.1. Spacetime diagrams representing (a) standing still and (b) moving to the right at constant speed.

5.3 LORENTZ TRANSFORMATIONS

We now relate Lorentz transformations, based on the physical postulates of special relativity, to hyperbola geometry. The Lorentz transformation between a frame (x, t) at rest and a frame (x', t') moving to the right at speed v was derived in Chapter 2. The transformation from the moving frame to the frame at rest is given by

$$x = \gamma (x' + vt'), \qquad (5.1)$$

$$t = \gamma \left(t' + \frac{v}{c^2} x'\right), \qquad (5.2)$$

where, as before,

$$\gamma = \frac{1}{\sqrt{1 - \frac{v^2}{c^2}}}. \qquad (5.3)$$

The key to converting this description to hyperbola geometry is to measure space and time in the same units by replacing t by ct. The transformation from the moving frame, which we now denote (x', ct'), to the frame at rest, now denoted (x, ct), is given by

$$x = \gamma (x' + \frac{v}{c} ct'), \qquad (5.4)$$

$$ct = \gamma \left(ct' + \frac{v}{c} x'\right), \qquad (5.5)$$

which makes the symmetry between these equations much more obvious.

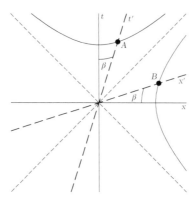

FIGURE 5.2. The Lorentz transformation between an observer at rest and an observer moving at speed $\frac{v}{c} = \tanh\beta$ is shown as a hyperbolic rotation. Point A has coordinates $(\rho \sinh\beta, \rho \cosh\beta)$, and units have been chosen such that $c = 1$.

We can simplify things still further by introducing the *rapidity* β via[3]

$$\frac{v}{c} = \tanh\beta. \tag{5.6}$$

By inserting Equation (5.6) into the expression for γ, (5.3), we obtain the identities

$$\gamma = \frac{1}{\sqrt{1 - \tanh^2\beta}} = \sqrt{\frac{\cosh^2\beta}{\cosh^2\beta - \sinh^2\beta}} = \cosh\beta \tag{5.7}$$

and

$$\frac{v}{c}\gamma = \tanh\beta \cosh\beta = \sinh\beta. \tag{5.8}$$

By then inserting identities (5.7) and (5.8) into the Lorentz transformations (5.4) and (5.5), we obtain the remarkably simple form

$$x = x' \cosh\beta + ct' \sinh\beta, \tag{5.9}$$
$$ct = x' \sinh\beta + ct' \cosh\beta, \tag{5.10}$$

which, in matrix form, is just

$$\begin{pmatrix} x \\ ct \end{pmatrix} = \begin{pmatrix} \cosh\beta & \sinh\beta \\ \sinh\beta & \cosh\beta \end{pmatrix} \begin{pmatrix} x' \\ ct' \end{pmatrix}. \tag{5.11}$$

But note that Equation (5.11) is just Equation (4.17), with $y = ct$.

---

[3] Warning: Some authors use β for $\frac{v}{c}$, not the rapidity.

Thus, Lorentz transformations are just hyperbolic rotations. As noted in Chapter 4, the invariance of the interval follows immediately from the fundamental hyperbolic trigonometric identity (4.4). This invariance now takes the form

$$x^2 - c^2 t^2 \equiv x'^2 - c^2 t'^2, \tag{5.12}$$

as shown in Figure 5.2, which is nearly identical to Figure 4.2. The physics of Lorentz transformations and the Lorentzian geometry of hyperbolas are therefore one and the same; hyperbola geometry is the geometry of special relativity.

5.4 SPACE AND TIME

We now return to the peculiar fact that the squared distance between two points can be positive, negative, or zero. The sign is positive for horizontal distances and negative for vertical distances. These directions correspond to the coordinates x and t and measure space and time, respectively—as seen by the given observer. But *any* observer's space axis must intersect the hyperbola $x^2 - c^2 t^2 = \rho^2$ somewhere, and hence corresponds to positive squared distance. Such directions have more space than time, and are called *spacelike*. Similarly, any observer's time axis intersects the hyperbola $x^2 - c^2 t^2 = -\rho^2$, corresponding to negative squared distance; such directions are called *timelike*.

What about diagonal lines at a Euclidean angle of 45°? These lines correspond to a squared distance of zero—and to moving at the speed of light. All observers agree about these directions, which are called *lightlike*. In hyperbola geometry, there are thus preferred directions of "length zero." Indeed, this is the geometric realization of the idea that the speed of light is the same for all observers.

It is important to realize that *every* spacelike direction corresponds to the space axis for *some* observer. Events separated by a spacelike line occur at the same time for that observer—and the square root of the squared distance is just the distance between the events as seen by that observer. Similarly, events separated by a timelike line occur at the same place for some observer, and the square root of the absolute value of the squared distance is just the time that elapses between the events as seen by that observer.

In contrast, events separated by a timelike line do not occur simultaneously for *any* observer. We can thus divide the spacetime diagram into

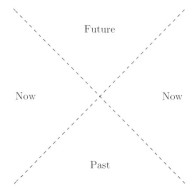

FIGURE 5.3. The causal relationship between points in spacetime and the origin.

causal regions as follows: those points connected to the origin by spacelike lines occur "now" for some observer, whereas those points connected to the origin by timelike lines occur unambiguously in the future or the past. This relationship is shown in Figure 5.3.[4]

To be able to make sense of cause and effect, only events in our past can influence us, and we can influence only events in our future. Put differently, if information could travel faster than the speed of light, then different observers would no longer be able to agree on cause and effect.

5.5 DOT PRODUCT

In Euclidean geometry, (squared) distances can be described by taking the dot product of a vector with itself. By denoting the unit vectors in the x and y directions by $\hat{\boldsymbol{x}}$ and $\hat{\boldsymbol{y}}$, respectively, we can write the vector from the origin to the point (x, y) as

$$\vec{r} = x\,\hat{\boldsymbol{x}} + y\,\hat{\boldsymbol{y}}, \tag{5.13}$$

whose squared length is

$$|\vec{r}|^2 = \vec{r} \cdot \vec{r} = x^2 + y^2. \tag{5.14}$$

---

[4]With two or more spatial dimensions, the lightlike directions would form a surface called the *light cone*, and the regions labeled "now" would be connected.

5.5. DOT PRODUCT

The generalization to hyperbola geometry is straightforward. We denote the unit vectors in the t and x directions by $\hat{t}$ and $\hat{x}$.[5] Then the Lorentzian dot product can be defined by the requirement that this be an orthonormal basis in the sense that

$$\hat{x} \cdot \hat{x} = 1, \tag{5.15}$$
$$\hat{t} \cdot \hat{t} = -1, \tag{5.16}$$
$$\hat{x} \cdot \hat{t} = 0. \tag{5.17}$$

Any point (x, ct) in spacetime can thus be identified with the vector

$$\vec{r} = x\,\hat{x} + ct\,\hat{t} \tag{5.18}$$

from the origin to that point, whose squared length is just the squared distance from the origin, namely,

$$|\vec{r}|^2 = \vec{r} \cdot \vec{r} = x^2 - c^2 t^2. \tag{5.19}$$

One of the fundamental properties of the Euclidean dot product is that

$$\vec{u} \cdot \vec{v} = |\vec{u}||\vec{v}| \cos\theta, \tag{5.20}$$

where θ is the smallest angle between the directions of $\vec{u}$ and $\vec{v}$. This relationship between the dot product and projections of one vector along another can in fact be used to *define* the dot product. But what happens in hyperbola geometry?

First of all, the dot product can be used to define right angles: Two vectors $\vec{u}$ and $\vec{v}$ are said to be *orthogonal* (or *perpendicular*) precisely when their dot product is zero:

$$\vec{u} \perp \vec{v} \iff \vec{u} \cdot \vec{v} = 0. \tag{5.21}$$

We adopt this definition unchanged in hyperbola geometry.

When are $\vec{u}$ and $\vec{v}$ perpendicular? Assume first that $\vec{u}$ is spacelike. Without loss of generality, let $\vec{u}$ be a unit vector, in which case it takes the form

$$\vec{u} = \cosh\alpha\,\hat{x} + \sinh\alpha\,\hat{t}. \tag{5.22}$$

Then one vector perpendicular to $\vec{u}$ would be

$$\vec{v} = \sinh\alpha\,\hat{x} + \cosh\alpha\,\hat{t}, \tag{5.23}$$

---

[5] Unit vectors are dimensionless. It is neither necessary nor desirable to include a factor of c in the definition of $\hat{t}$.

and it is easy to check that all other vectors perpendicular to $\vec{u}$ are multiples of this one. Note that $\vec{v}$ is timelike. Had we assumed instead that $\vec{u}$ were timelike, we would merely have interchanged the roles of $\vec{u}$ and $\vec{v}$.

Furthermore, $\vec{u}$ and $\vec{v}$ are just the space and time axes, respectively, of an observer moving with speed $\frac{v}{c} = \tanh \alpha$. So orthogonal directions correspond precisely to the coordinate axes of some observer.

What if $\vec{u}$ is lightlike? It is a peculiarity of Lorentzian hyperbola geometry that there are nonzero vectors of length zero. But since the dot product of a vector with itself gives its squared length, having length zero means that lightlike vectors are perpendicular to themselves.

We can finally define the *length* of a vector $\vec{v}$ by

$$|\vec{v}| = \sqrt{|\vec{v} \cdot \vec{v}|}. \tag{5.24}$$

If $\vec{v}$ is spacelike, we can write

$$\vec{v} = \pm |\vec{v}|(\cosh \alpha \, \hat{x} + \sinh \alpha \, \hat{t}), \tag{5.25}$$

whereas if $\vec{v}$ is timelike, we can write

$$\vec{v} = \pm |\vec{v}|(\sinh \alpha \, \hat{x} + \cosh \alpha \, \hat{t}). \tag{5.26}$$

If $\vec{v}$ is lightlike, however, $|\vec{v}| = 0$, so no such expression exists. Rather,

$$\vec{v} = a\,(\hat{x} \pm \hat{t}) \tag{5.27}$$

for some constant a.

The above argument shows that timelike vectors can only be perpendicular to spacelike vectors, and vice versa. We also say in this case that the vectors form a *right angle*. Recall that hyperbolic angles are defined along hyperbolas of the form $x^2 - c^2 t^2 = \rho^2$, and hence exist (as originally defined) only between spacelike directions. It is straightforward to extend this to timelike directions by using the hyperbola $x^2 - c^2 t^2 = -\rho^2$; this was implicitly done in Figure 4.2. But there is no hyperbola relating timelike directions to spacelike ones. Thus, a "right angle" isn't an angle at all!

A *right triangle* is one that contains a right angle. By the above discussion, one of the legs of such a triangle must be spacelike and the other timelike. Consider first the case where the hypotenuse is either spacelike or timelike. The only hyperbolic angle in such a triangle is the one between the hypotenuse and the leg of the same type, that is, between the two time-

5.5. Dot Product

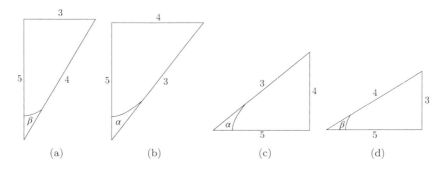

FIGURE 5.4. Some hyperbolic right triangles.

like sides if the hypotenuse is timelike and between the two spacelike sides if the hypotenuse is spacelike. Several such hyperbolic right triangles are shown in Figure 5.4(a)–(d). It is also possible for the hypotenuse to be null, as shown in Figure 5.5; such triangles do not have any hyperbolic angles. Finally, as seen in Figures 5.5(b) and 5.5(c), right angles do not always look like right angles. The reader is encouraged to combine these ideas and attempt to draw a 3–4–5 triangle, none of whose sides are horizontal or vertical.

What happens if we take the dot product between two spacelike vectors? We can assume without loss of generality that one vector is parallel to the x-axis, in which case we have

$$\vec{u} = |\vec{u}|\,\hat{x}, \tag{5.28}$$
$$\vec{v} = |\vec{v}|(\cosh\alpha\,\hat{x} + \sinh\alpha\,\hat{t}), \tag{5.29}$$

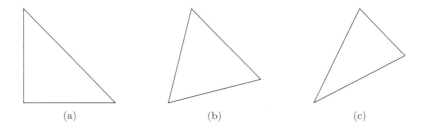

FIGURE 5.5. More hyperbolic right triangles. The right angle is on the left.

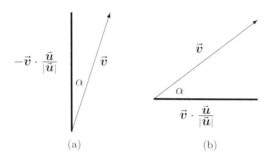

FIGURE 5.6. Hyperbolic projections (a) between two spacelike vectors or (b) between two timelike vectors. In both cases, $\vec{u}$ points along the heavy line.

so that the dot product satisfies

$$\vec{u} \cdot \vec{v} = |\vec{u}||\vec{v}| \cosh \alpha. \tag{5.30}$$

What happens if both vectors are timelike? The above argument still works, except that the roles of $\hat{x}$ and $\hat{t}$ must be interchanged, resulting in

$$\vec{u} \cdot \vec{v} = -|\vec{u}||\vec{v}| \cosh \alpha. \tag{5.31}$$

In both cases, note that $|\vec{v}| \cosh \alpha$ is the projection of $\vec{v}$ along $\vec{u}$; see Figure 5.6.

What happens if we take the dot product between a timelike vector and a spacelike vector? We can again assume without loss of generality that the spacelike vector is parallel to the x-axis, so that

$$\vec{u} = |\vec{u}| \hat{x}, \tag{5.32}$$
$$\vec{v} = |\vec{v}|(\sinh \alpha \, \hat{x} + \cosh \alpha \, \hat{t}). \tag{5.33}$$

The dot product now satisfies

$$\vec{u} \cdot \vec{v} = |\vec{u}||\vec{v}| \sinh \alpha. \tag{5.34}$$

At first sight, this is something new. But note from Figure 5.7(a) that $\vec{v} \sinh \alpha$ is just the projection of $\vec{v}$ along $\vec{u}$! The new feature here is that

5.5. DOT PRODUCT

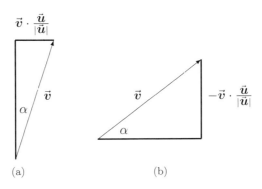

FIGURE 5.7. Hyperbolic projections between (a) timelike and (b) spacelike vectors. The orientation of $\vec{u}$ in each case is the opposite of that shown in Figure 5.6, namely, horizontally in (a) and vertically in (b).

we cannot define the angle between a timelike direction and a spacelike direction. The only angle in the triangle that is defined is the one shown.[6]

---

[6] Alternatively, we could have assumed that the timelike vector was parallel to the t-axis, resulting in Figure 5.7(b). The conclusion is the same, although now it represents the projection of $\vec{u}$ along $\vec{v}$.

➤ Chapter 6 ➤

Applications

In which we learn some of the consequences of special relativity.

6.1 Drawing Spacetime Diagrams

We begin by summarizing the rules for drawing spacetime diagrams.

- Points in spacetime are called *events*.
- Lines with slope $m = \pm 1$ represent beams of light.
- Vertical lines represent the worldline of an object at rest.
- Horizontal lines represent snapshots of constant time, that is, events that are simultaneous (in the given reference frame).
- Lines with slope $|m| > 1$ (called *timelike*) represent the worldlines of observers moving at constant speed.
- The speed of such observers is given by $c \tanh\beta$, where β is the (hyperbolic) angle between the worldline and a *vertical* line.
- The distance between two events along a timelike line is just the time between them as measured by the moving observer.
- Lines with slope $|m| < 1$ (called *spacelike*) represent lines of simultaneity as seen by an observer moving at constant speed.
- The speed of this observer is given by $c \tanh\beta$, where β is the (hyperbolic) angle between the line of simultaneity and a *horizontal* line.
- The distance between two events along a spacelike line is just the distance between them as measured by this observer.

- Hyperbolas centered at the origin represent events at constant "distance" from the origin; such hyperbolas can be used to calibrate the scales along any line.

- Lines are orthogonal if they have reciprocal slopes.

- Triangle trigonometry can be used to relate measurements in different reference frames. *(Be careful: a right triangle contains only one angle.)*

6.2 Addition of Velocities

What does the rapidity β represent? Consider an observer moving at speed v to the right. This observer's world line intersects the hyperbola

$$c^2 t^2 - x^2 = \rho^2 \qquad (ct > 0) \tag{6.1}$$

at the point A with coordinates $(\rho \sinh \beta, \rho \cosh \beta)$; this line therefore has "slope"[1]

$$\frac{v}{c} = \tanh \beta, \tag{6.2}$$

which is the same as Equation (5.6). Thus, β is nothing more than the *hyperbolic angle* between the ct-axis and the worldline of a moving object. As discussed in Chapter 4, β is precisely the distance from the axis as measured along the hyperbola (in hyperbola geometry). This relationship was illustrated in Figures 4.2 and 5.2.

Consider, therefore, an object moving at speed u relative to an observer moving at speed v. Their rapidities are given by, respectively,

$$\frac{u}{c} = \tanh \alpha, \tag{6.3}$$

$$\frac{v}{c} = \tanh \beta. \tag{6.4}$$

To determine the resulting speed with respect to an observer at rest, we simply add the *rapidities*! One way to think of this is that we are adding the arc lengths along the hyperbola. Another is that we are following a

---

[1] It is not obvious whether slope should be defined by $\frac{\Delta x}{c \Delta t}$ or by the reciprocal of this expression. This is further complicated by the fact that both (x, ct) and (ct, x) are commonly used to denote the coordinates of the point A.

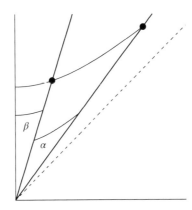

FIGURE 6.1. The geometry of the Einstein addition formula.

hyperbolic rotation through a hyperbolic angle β (to get to the moving observer's frame) with a rotation through an angle α, as shown in Figure 6.1. In any case, the resulting speed w is given by

$$\frac{w}{c} = \tanh(\alpha + \beta) = \frac{\tanh \alpha + \tanh \beta}{1 + \tanh \alpha \tanh \beta} = \frac{\frac{u}{c} + \frac{v}{c}}{1 + \frac{uv}{c^2}}, \qquad (6.5)$$

which is—finally—precisely the Einstein addition formula.

6.3 LENGTH CONTRACTION

We now return to the question of how "wide" things are. Consider first a meter stick at rest. In spacetime, the stick "moves" vertically, that is, it ages. This situation is shown in Figure 6.2(a), where the horizontal lines show the meter stick at various times (according to an observer at rest). How "wide" is the *worldsheet* of the stick? The observer at rest, of course, measures the length of the stick by locating both ends *at the same time* and measuring the distance between them. At $t = 0$, this corresponds to the two heavy dots in the sketch, one at the origin and the other on the unit hyperbola. But *all* points on the unit hyperbola are at an interval of 1 meter from the origin. The observer at rest therefore concludes, unsurprisingly, that the meter stick is 1 meter long.

How long does a moving observer think the stick is? This length is just the width of the worldsheet *as measured by the moving observer*. This

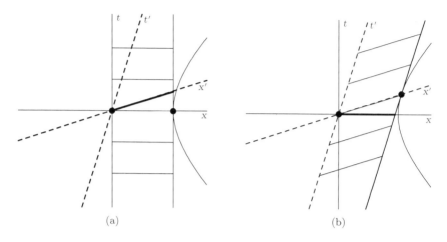

FIGURE 6.2. Length contraction as a hyperbolic projection: (a) object at rest and observer in motion; (b) object in motion and observer at rest.

observer follows the same procedure, by locating both ends of the stick *at the same time*, and measuring the distance between them. But time now corresponds to t', not t. At $t' = 0$, this measurement corresponds to the heavy line in Figure 6.2(a). Since this line fails to reach the unit hyperbola, it is clear that the moving observer measures the length of a stationary meter stick to be less than 1 meter. This is length contraction.

To determine the exact value measured by the moving observer, it is necessary only to note that the triangle shown in Figure 6.2(a) is a hyperbolic right triangle, with the heavy line indicating the hypotenuse. Simple hyperbolic trigonometry shows immediately that the hypotenuse of a right triangle with angle β and adjacent side of length 1 is given by $1/\cosh\beta$.

For those who prefer an algebraic derivation, we compute the intersection of the line $x = 1$ (the right-hand edge of the meter stick) with the line $t' = 0$ (the x'-axis), or, equivalently, $ct = x\tanh\beta$, to find that

$$ct = \tanh\beta, \tag{6.6}$$

so that x' is just the interval from this point to the origin, which is

$$x' = \sqrt{x^2 - c^2 t^2} = \sqrt{1 - \tanh^2\beta} = \frac{1}{\cosh\beta}. \tag{6.7}$$

What if the stick is moving and the observer is at rest? This situation is shown in Figure 6.2(b). The worldsheet now corresponds to a "rotated rectangle," indicated by the parallelograms in the sketch. The fact that the meter stick is 1 meter long in the moving frame is shown by the distance between the two heavy dots (along $t' = 0$), and the measurement by the observer at rest is indicated by the heavy line (along $t = 0$). Again, it is clear that the stick appears to have shrunk, since the heavy line fails to reach the unit hyperbola. To determine the exact value measured by the observer at rest, proceed as before using the hyperbolic right triangle shown in Figure 6.2(b), which is however congruent to the triangle shown in Figure 6.2(a), so that the result is the same.

Thus, a moving object appears shorter by a factor of $1/\cosh\beta$. It doesn't matter whether it is the stick or the observer that is moving; all that matters is their relative motion.

6.4 TIME DILATION

We now investigate moving clocks. Consider first the smaller dot in Figure 6.3. This dot corresponds to $ct = 1$ (and $x = 0$), as evidenced by the fact that this point is on the other unit hyperbola, as shown. Similarly, the larger dot, lying on the same hyperbola, corresponds to $ct' = 1$ (and $x' = 0$). The horizontal line emanating from this dot gives the value of ct there, which is clearly greater than 1 and which represents the time according to the observer at rest when the moving clock says 1. According to the observer at rest, the moving clock therefore runs slow. But now consider

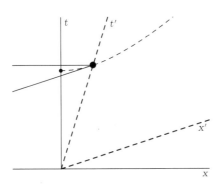

FIGURE 6.3. Time dilation as a hyperbolic projection.

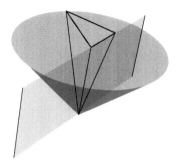

FIGURE 6.4. The three-dimensional spacetime diagram for a bouncing beam of light on a moving train. The beam of light is along the light cone, and is projected into both the xy-plane (top) and the xt-plane (front). (See also page II.)

the diagonal line emanating from the larger dot. At all points along this line, $ct' = 1$. In particular, at the smaller dot we must have $ct' > 1$. Thus, the time according to the moving observer when the clock at rest says 1 (at the smaller dot) must be greater than 1; the moving observer therefore concludes the clock at rest runs slow.

There is no contradiction here; we must simply be careful to ask the right question. In each case, observing a clock in another frame of reference corresponds to a projection. In each case, a clock in relative motion to the observer appears to run slow.

To determine the exact value measured by the moving observer, we return to the bouncing beam of light considered in Chapter 2. Figure 2.5 showed the relationships between the various *distances*, but we would now like to draw a spacetime diagram for this scenario. However, since the beam of light is moving in both the horizontal (x) and vertical (y) directions, we need a *three-dimensional* spacetime diagram, as shown in Figure 6.4.

In Figure 6.4, the vertical line represents an observer at rest on the platform, and the mostly vertical line on the right represents the motion of the flashlight on the floor of the moving train. The beam of light is the line in the back of the diagram, along the light cone, and moving in both the x and y directions.

By projecting Figure 6.4 into a horizontal plane (with $t = $ constant), as shown at the top of the cone, we recover precisely Figure 2.5. The base of the triangle corresponds to the distance $v \Delta t$ traveled by the train, the hypotenuse corresponds to the distance $c \Delta t$ traveled by the beam of light according to the observer on the platform, and the remaining leg of the

6.4. TIME DILATION

FIGURE 6.5. The spacetime Pythagorean theorem for a bouncing light beam on a moving train.

triangle corresponds to the height of the train, that is, to the distance $c\,\Delta t'$ traveled by the beam of light according to the observer on the train. Projecting Figure 6.4 instead into the vertical plane (with $y = 0$), as shown in the middle of the cone, we obtain the spacetime diagram shown in Figure 6.5.

Remarkably, the edges of the two triangles formed by the horizontal and vertical projections have the same lengths. To see this, we note first of all that they share one edge. But each of the remaining pairs of edges form the legs of a right triangle whose hypotenuse is the beam of light. The speed of light is 1 (in appropriate units), so in each case the legs must have the same length. This argument justifies the labeling used in Figure 6.5, and can in fact be used to *derive* the hyperbolic Pythagorean theorem, which says here that
$$(c\,\Delta t')^2 = (c\,\Delta t)^2 - (v\,\Delta t)^2, \tag{6.8}$$
or, equivalently, that
$$t' = \frac{t}{\cosh \beta}, \tag{6.9}$$
where $\tanh \beta = v/c$ is the speed of the train. Moving clocks therefore run slow, by a factor of precisely $\cosh \beta$.

It is worth comparing Figures 2.5 and 6.5, using Figure 6.4 as needed. These two diagrams are almost the same, but two of the edges appear to have been interchanged. In Figure 2.5, the vertical *distance* traveled by the beam of light was $c\Delta t'$; in Figure 6.5, the *time* taken by the beam to reach the top of the train is $c\Delta t'$ (in appropriate units), namely, the length of the projected worldline of the beam of light, which is the hypotenuse of the triangle.

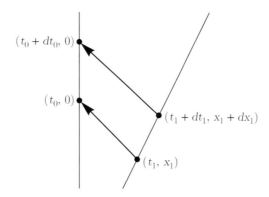

FIGURE 6.6. The Doppler effect. An observer moving to the right emits a pulse of light to the left, which is later seen by a stationary observer. The wavelengths measured by the two observers differ, causing a *Doppler shift* in the frequency.

6.5 DOPPLER SHIFT

The frequency f of a beam of light is related to its wavelength λ by the formula

$$f\lambda = c. \tag{6.10}$$

How do these quantities depend on the observer?

Consider an inertial observer moving to the right in the laboratory frame and carrying a flashlight that is pointing to the left; see Figure 6.6. Then the moving observer is traveling along a path of the form $x' = x'_1 =$ constant. Suppose the moving observer turns on the flashlight (at time t'_1) just long enough to emit one complete wavelength of light, and that this takes time dt'. Then the moving observer "sees" a wavelength

$$\lambda' = c\, dt'. \tag{6.11}$$

According to an observer in the lab, the flashlight was turned on at the event (t_1, x_1) and turned off dt_1 seconds later, during which time the moving observer moved a distance dx_1 meters to the right. But when was the light received, at, say, $x = 0$?

Let $(t_0, 0)$ denote the first reception of light by a lab observer at $x = 0$, and suppose this observer sees the light stay on for dt_0 seconds. Since light travels at the speed of light, we have

$$c(t_0 - t_1) = x_1 \tag{6.12}$$

6.5. Doppler Shift

and
$$c[(t_0 + dt_0) - (t_1 + dt_1)] = x_1 + dx_1, \tag{6.13}$$

from which it follows that
$$c(dt_0 - dt_1) = dx_1, \tag{6.14}$$

so that
$$\begin{aligned} c\,dt_0 &= dx_1 + c\,dt_1 \\ &= (dx'_1 \cosh\beta + c\,dt'_1 \sinh\beta) + (c\,dt'_1 \cosh\beta + dx'_1 \sinh\beta) \\ &= (\cosh\beta + \sinh\beta)\,c\,dt'_1, \end{aligned} \tag{6.15}$$

since $dx'_1 = 0$. But the wavelength as seen in the lab is
$$\lambda = c\,dt_0, \tag{6.16}$$

so that
$$\begin{aligned} \frac{\lambda}{\lambda'} &= \frac{dt_0}{dt'_1} = \cosh\beta + \sinh\beta \\ &= \cosh\beta\,(1 + \tanh\beta) = \gamma\left(1 + \frac{v}{c}\right) = \sqrt{\frac{1 + \frac{v}{c}}{1 - \frac{v}{c}}}. \end{aligned} \tag{6.17}$$

The frequencies transform inversely, that is,
$$\frac{f'}{f} = \sqrt{\frac{1 + \frac{v}{c}}{1 - \frac{v}{c}}}. \tag{6.18}$$

⪼ Chapter 7 ⪻

Problems I

7.1 Practice

PROBLEM 7.1. *In a laboratory experiment a muon is observed to travel 800 meters before disintegrating. The lifetime of a muon is 2×10^{-6} seconds, so the speed must be*

$$v = \frac{800 \ m}{2 \times 10^{-6} \ s} = 4 \times 10^8 \ m/s, \tag{7.1}$$

which is faster than the speed of light. Identify the error in this computation, and find the actual speed of the muon.

The lifetime of the muon must be measured in its own rest frame. Geometrically, we have the triangle in Figure 7.1, the *hypotenuse* of which is

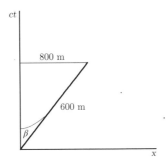

FIGURE 7.1. The spacetime diagram for a traveling muon.

$2c \times 10^{-6} = 600$ meters. The remaining leg of this 3–4–5 triangle is easily seen to be 1000 meters, from which we can read off

$$\frac{v}{c} = \tanh \beta = \frac{4}{5}. \tag{7.2}$$

PROBLEM 7.2. *A rocket ship leaves Earth at a speed of $\frac{3}{5}c$. When a clock on the rocket says 1 hour has elapsed, the rocket sends a light signal back to Earth.*

(a) *According to Earth clocks, when was the signal sent?*

(b) *According to Earth clocks, how long after the rocket left did the signal arrive back on Earth?*

(c) *According to the rocket observer, how long after the rocket left did the signal arrive back on Earth?*

All of these questions can be answered by drawing a spacetime diagram, as shown in Figure 7.2. We again have a 3–4–5 triangle, with hypotenuse of length 1 and $\tanh \beta = \frac{3}{5}$. We therefore have $\cosh \beta = \frac{5}{4}$ (the vertical leg) and $\sinh \beta = \frac{3}{4}$ (the horizontal leg).

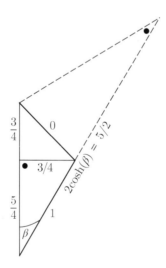

FIGURE 7.2. The spacetime diagram for the rocket ship. The dots indicate right angles.

7.1. PRACTICE

(a) The signal was sent after $\frac{5}{4}$ hours according to Earth clocks.

(b) The signal was received $\frac{3}{4}$ hours after it was sent, or two hours after the rocket left, again according to Earth clocks.

(c) We consider the large right triangle in Figure 7.2, also with hyperbolic angle β, and with vertical side 2. This side is the hypotenuse, since the right angle is in the upper right corner. We are trying to find the remaining side, which is $2\cosh\beta = \frac{5}{2}$, so the signal reaches Earth $\frac{5}{2}$ hours after the rocket left, according to its own clocks.

PROBLEM 7.3. *A Lincoln Continental is twice as long as a VW Beetle when they are at rest. As the Lincoln overtakes the VW, going through a speed trap, a stationary policeman observes that they both have the same length. The VW is going at half the speed of light. How fast is the Lincoln going?*

Suppose the speed (really the rapidity) of the VW is given by $\tanh\alpha$ and that of the Lincoln by $\tanh\beta$. A spacetime diagram showing the worldlines of the front and back of each car is shown in Figure 7.3(a). Since the cars appear to be the same length, these worldlines intersect along the horizontal axis. The "true" lengths of the cars, namely L and $2L$, respectively, are also shown. Comparing the right triangles shown, whose right angles are indicated with dots and which share a common horizontal leg, we see that

$$\frac{L}{\cosh\alpha} = \frac{2L}{\cosh\beta}, \tag{7.3}$$

so that

$$\cosh\beta = 2\cosh\alpha. \tag{7.4}$$

But we are given that $\tanh\alpha = \frac{1}{2}$, so that, by using the triangles shown in Figure 7.3(b), we have in turn that

$$\cosh\alpha = \frac{2}{\sqrt{3}}, \tag{7.5}$$

$$\cosh\beta = \frac{4}{\sqrt{3}}, \tag{7.6}$$

$$\tanh\beta = \frac{\sqrt{13}}{4}, \tag{7.7}$$

so that the Lincoln is moving at $\frac{\sqrt{13}}{4}$ the speed of light.

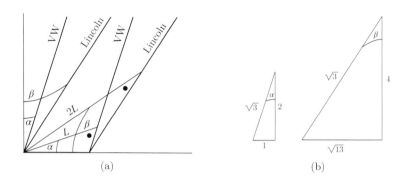

FIGURE 7.3. (a) The spacetime diagram for the Lincoln and the VW (not to scale). (b) Two helpful hyperbolic right triangles.

PROBLEM 7.4. *Sophie Zabar, clairvoyante, cried out in pain at precisely the instant her twin brother, 500 km away, hit his thumb with a hammer. A skeptical scientist observed both events from an airplane traveling at $\frac{12}{13}c$ to the right. Which event occurred first, according to the scientist? How much earlier was it?*

This situation is shown in Figure 7.4, from the shared reference frame of the clairvoyante (C) and her brother (B). The worldline of the scientist (S) is tilted to the right with hyperbolic angle β; his lines of simultaneity make

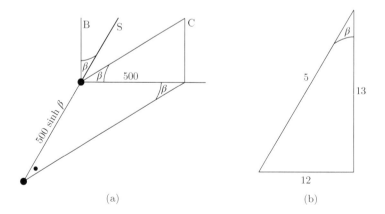

FIGURE 7.4. (a) Spacetime diagram for the scientist's analysis (not to scale). (b) Helpful right triangle.

7.2. THE GETAWAY

the same angle with the horizontal axis. His time is measured along his own worldline, so we are looking for the "distance" between the two heavy dots. Triangle trigonometry does the rest; this leg has length in kilometers given by

$$500 \sinh \beta = 1200, \tag{7.8}$$

where we have made use of the smaller 5–12–13 triangle shown in the figure to compute $\sinh \beta = \frac{12}{5}$ from $\tanh \beta = \frac{12}{13}$. Converting this distance to a time by dividing by c, we conclude that the scientist observes the brother hit his thumb 4×10^{-3} seconds *before* Sophie cried out.

7.2 THE GETAWAY

PROBLEM 7.5. *The outlaws are escaping in their getaway car, which moves at $\frac{3}{4}c$, chased by the police, moving at only $\frac{1}{2}c$. Realizing they can't catch up, the police attempt to shoot out the tires of the getaway car. Their guns have a muzzle velocity (speed of the bullets relative to the gun) of $\frac{1}{3}c$.*

(a) Does the bullet reach its target according to Galileo?

(b) Does the bullet reach its target according to Einstein?

(c) Verify that your answer to part (b) is the same in all four reference frames: ground, police, outlaws, and bullet.

(a) In Newtonian physics, we add the speed of the bullet ($\frac{1}{3}c$) to that of the police ($\frac{1}{2}c$) to compute the speed of the bullet with respect to the ground, obtaining $\frac{5}{6}c$, which is greater than the speed of the outlaws ($\frac{3}{4}c$). Yes, the bullet reaches its target according to Galileo.

(b) Denote the rapidities of the police and outlaws with respect to the ground by α and β, respectively, and let the rapidity of the bullet with respect to the police be γ. Then we have

$$\tanh \alpha = \frac{1}{2}, \tag{7.9}$$

$$\tanh \beta = \frac{3}{4}, \tag{7.10}$$

$$\tanh \gamma = \frac{1}{3}. \tag{7.11}$$

	Ground	Police	Outlaws	Bullet
Ground	0	$\frac{1}{2}$	$\frac{3}{4}$	$\frac{5}{7}$
Police	$-\frac{1}{2}$	0	$\frac{2}{5}$	$\frac{1}{3}$
Outlaws	$-\frac{3}{4}$	$-\frac{2}{5}$	0	$-\frac{1}{13}$
Bullet	$-\frac{5}{7}$	$-\frac{1}{3}$	$\frac{1}{13}$	0

TABLE 7.5. The relative speeds of the ground, police, outlaws, and bullet with respect to each other.

In special relativity, we add the rapidities (hyperbolic angles), so that the speed of the bullet with respect to the ground is given by

$$\frac{v_b}{c} = \tanh(\alpha + \gamma) = \frac{\tanh\alpha + \tanh\gamma}{1 + \tanh\alpha \tanh\gamma} = \frac{5}{7}. \quad (7.12)$$

Thus, v_b is less than the speed of the outlaws ($\frac{3}{4}c$), and the bullet does not reach its target, according to Einstein.

(c) By performing a similar computation in each case, we obtain Table 7.5, in which all speeds are given as fractions of c.

7.3 Angles Are Not Invariant

PROBLEM 7.6. *The mast of a sailboat leans at an angle θ (measured from the deck) toward the rear of the boat.*

(a) *An observer on the dock sees the boat go by at speed $v \ll c$ (so you do not need to use relativity to do this problem). What angle does the observer say the mast makes?*

(b) *A child on the boat throws a ball into the air at the same angle θ. What angle does the observer on the dock say the ball makes with the deck? (Ignore the subsequent influence of gravity on the ball—this question is only about the initial angle when the ball leaves the child's hand.)*

7.3. ANGLES ARE NOT INVARIANT

FIGURE 7.6. (a) The angle made by the mast of a moving sailboat. (b) The speed of the mast. (Primed coordinates refer to the dock.)

(c) A spaceship goes past the dock at speed v. An antenna is mounted on its hull at an angle θ with the (horizontal) hull. What angle does the observer on the dock say the antenna makes?

(d) A spotlight is mounted on the spaceship so that its beam makes an angle θ with the hull. What angle does the observer on the dock say the beam makes with the hull?

(a) See Figure 7.6. The angles are the same, since both observers agree on the lengths of the sides of the right triangle made by the mast and the deck. Using primed coordinates to refer to the dock, and unprimed coordinates for the sailboat, we have

$$\Delta x' = \Delta x, \tag{7.13}$$
$$\Delta y' = \Delta y, \tag{7.14}$$

so that

$$\tan \theta' = \frac{\Delta y'}{\Delta x'} = \frac{\Delta y}{\Delta x} = \tan \theta. \tag{7.15}$$

(b) See Figure 7.7. The right triangle used to determine this angle now involves velocities, not lengths, and velocities are different as measured by the two observers. We now have

$$u'_x = u_x - v, \tag{7.16}$$
$$u'_y = u_y, \tag{7.17}$$

so that

$$\tan \theta' = \frac{u'_y}{u'_x} = \frac{u_y}{u_x - v} = \frac{\tan \theta}{1 - \frac{v}{u} \cos \theta} = \frac{u \sin \theta}{u \cos \theta - v}. \tag{7.18}$$

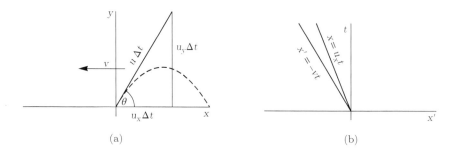

FIGURE 7.7. (a) The angle made by a ball thrown from a moving sailboat. (b) The speed of the ball.

(c) See Figure 7.8. Now we have to take length contraction into account. Thus,

$$\Delta x' = \frac{\Delta x}{\cosh \beta}, \qquad (7.19)$$
$$\Delta y' = \Delta y, \qquad (7.20)$$

so that

$$\tan \theta' = \frac{\Delta y'}{\Delta x'} = \frac{\Delta y}{\Delta x} \cosh \beta = \tan \theta \cosh \beta > \tan \theta. \qquad (7.21)$$

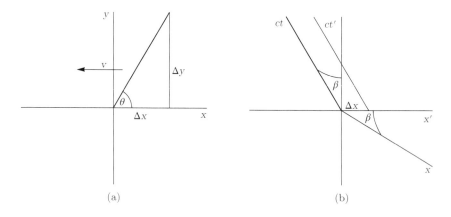

FIGURE 7.8. (a) The angle made by an antenna on a moving spaceship. (b) The Lorentz transformation between frames.

7.4. INTERSTELLAR TRAVEL

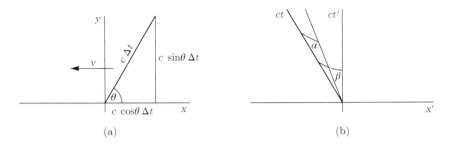

FIGURE 7.9. (a) The (Euclidean) angle made by the beam of a moving searchlight. (b) The (hyperbolic) angle between the two reference frames.

(d) See Figure 7.9. We must now deal with both the hyperbolic angle due to the spaceship's velocity and the Euclidean angle formed by the beam of light, which does *not* travel at the speed of light *in the direction of the spaceship*. We have, as usual, $\frac{v}{c} = \tanh\beta$, where v is the speed of the spaceship. The component of the velocity of the beam of light *in this direction* is given by

$$\tanh\alpha = \cos\theta, \qquad (7.22)$$

as can be seen by using Figure 7.9(a). By using Figure 7.9(b), we see that

$$-\cos\theta' = \tanh(\beta - \alpha) = \frac{\tanh\beta - \tanh\alpha}{1 - \tanh\beta\tanh\alpha} = \frac{v - c\cos\theta}{c - v\cos\theta}. \qquad (7.23)$$

7.4 INTERSTELLAR TRAVEL

PROBLEM 7.7. *Alpha Centauri is roughly 4 lightyears from Earth. Dr. X travels (at constant velocity) from Earth to Alpha Centauri in 3 years. Immediately upon her arrival at Alpha Centauri, she turns on a powerful laser aimed at Earth. (Ignore Earth's motion.)*

(a) *How fast did Dr. X travel?*

(b) *How far did Dr. X travel? (That is, how far does she think Alpha Centauri is from Earth?)*

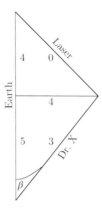

FIGURE 7.10. The spacetime diagram for Dr. X.

(c) *How long after Dr. X's departure do observers on Earth find out she arrived safely?*

(d) *Draw a spacetime diagram showing the worldlines of Dr. X, Earth, and the laser beam. Clearly indicate which events correspond to departure, arrival/turning on the laser, and the receipt of the laser signal on Earth.*

See the spacetime diagram in Figure 7.10, which answers part (d).

(a) The known data are that Alpha Centauri is 4 lightyears from Earth and that Dr. X travels for 3 years *as measured by her own clock*. The lower triangle in Figure 7.10 therefore has horizontal leg $\Delta x = 4$ lightyears and hypotenuse $c\Delta \tau = 3$ lightyears. The remaining (vertical) leg of this triangle therefore has length 5 lightyears, from which we can read off

$$\frac{v}{c} = \tanh\beta = \frac{4}{5}. \tag{7.24}$$

(b) This same triangle gives us a Lorentz contraction factor of

$$\cosh\beta = \frac{5}{3}, \tag{7.25}$$

from which it follows that Dr. X thinks that the distance from Earth to Alpha Centauri is

$$\ell' = \frac{\ell}{\cosh\beta} = \frac{4}{\frac{5}{3}} = \frac{12}{5}. \tag{7.26}$$

Alternatively, if Dr. X travels for 3 years at $\frac{4}{5}$ the speed of light, she must have traveled $\frac{12}{5}$ lightyears.

(c) We already know that the vertical leg of the lower triangle has length 5, indicating that observers on Earth believe Dr. X arrives 5 years after departure. The laser signal will take 4 years to travel 4 lightyears, for a total time difference of 9 years.

7.5 COSMIC RAYS

PROBLEM 7.8. *Consider muons produced by the collision of cosmic rays with gas nuclei in the atmosphere 60 kilometers above the surface of the earth, which then move vertically downward at nearly the speed of light. The half-life before muons decay into other particles is 1.5 microseconds (1.5×10^{-6} s).*

(a) Assuming it doesn't decay, how long would it take a muon to reach the surface of the earth?

(b) Assuming there were no time dilation, approximately what fraction of the muons would reach the earth without decaying?

(c) In actual fact, roughly $\frac{1}{8}$ of the muons would reach the earth. How fast are they going?

(a) Without much loss of accuracy, assume the muons travel at the speed of light. Then it takes them

$$\frac{60 \text{ km}}{3 \times 10^8 \text{ m/s}} = 200 \text{ }\mu\text{s} \tag{7.27}$$

to reach the earth.

(b) Since 200 μs is $\frac{200}{1.5} = \frac{400}{3}$ half-lives, only $2^{-\frac{400}{3}}$ of the muons reach the earth.

(c) *First solution:* The given fraction corresponds to 3 half-lives, since $\frac{1}{8} = 2^{-3}$. Thus, the time is dilated by a factor of $\frac{400/3}{3}$, so that

$$\cosh \alpha = \frac{400}{9}. \tag{7.28}$$

But

$$\frac{v}{c} = \tanh \alpha = \frac{\sqrt{400^2 - 9^2}}{400} \approx .99974684. \tag{7.29}$$

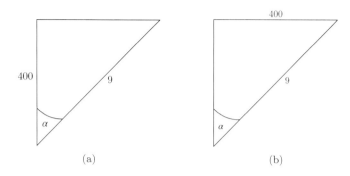

FIGURE 7.11. Hyperbolic triangles for the cosmic ray example. (a) Diagram for first solution. (b) Diagram for second solution.

See Figure 7.11(a).

Second solution: The first solution argument assumes $v \approx c$. (This was used to obtain the figure 200 μs in part (a), which is hence only an approximation.)

A more accurate argument would use the fact that the muons travel 60 km in $3 \times 1.5 \times 10^{-6}$ s (3 half-lives) *of proper time*. Thus,

$$\sinh \alpha = \frac{(60 \text{ km})(1000 \text{ m/km})}{(4.5 \times 10^{-6} \text{ s})(3 \times 10^8 \text{ m/s})} = \frac{400}{9}, \quad (7.30)$$

so that

$$\frac{v}{c} = \tanh \alpha = \frac{400}{\sqrt{400^2 + 9^2}} \approx .99974697. \quad (7.31)$$

See Figure 7.11(b).

It is important to realize not only that the second solution is more accurate than the first (assuming sufficient accuracy in the original data), but also that that the "shortcut" used in the first solution is justified.[1]

---

[1] This can be made rigorous by using a power series expansion. Equivalently, by simply redoing the computation using the approximation obtained in part (c) to recalculate part (a), the correct answer can be obtained to any desired accuracy. The speed at which this iterative procedure converges to the exact answer justifies having made the approximation in the first place.

7.6. DOPPLER EFFECT

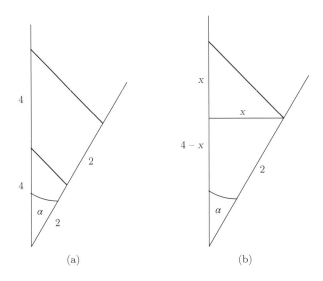

FIGURE 7.12. Computing Doppler shift. (a) Two flashes of light. (b) An enlarged view showing just one flash of light.

7.6 DOPPLER EFFECT

PROBLEM 7.9. *A rocket sends out flashes of light every two seconds in its own rest frame, which you receive every 4 seconds. How fast is the rocket going?*

First solution: This situation is shown in Figure 7.12(a). To find the hyperbolic angle α, draw a horizontal line as shown in the enlargement in Figure 7.12(b), resulting in the system of equations[2]

$$\tanh \alpha = \frac{x}{4-x}, \tag{7.33}$$

$$(4-x)^2 - x^2 = 2^2, \tag{7.34}$$

which is easily solved for $x = \frac{3}{2}$, so that $\frac{v}{c} = \tanh \alpha = \frac{3}{5}$.

Second solution: Insert $\lambda = 4$ and $\lambda' = 2$ into Equation (6.17) and solve for $\frac{v}{c}$.

---

[2] This method can be be used to derive the Doppler shift formula in general, yielding

$$\frac{v}{c} = \frac{\lambda^2 - \lambda'^2}{\lambda^2 + \lambda'^2}, \tag{7.32}$$

which is equivalent to Equation (6.17); in this example, $\lambda = 4$ and $\lambda' = 2$.

≻ Chapter 8 ≺

Paradoxes

In which impossible things are shown to be possible.

8.1 Special Relativity Paradoxes

It is easy to create seemingly impossible scenarios in special relativity by playing on the counterintuitive nature of observer-dependent time. These scenarios are usually called paradoxes because they seem to be impossible. Yet there is nothing paradoxical about them.

The best way to resolve these paradoxes is to draw a good spacetime diagram. This requires careful reading of the problem, making sure always to associate the given information with a particular reference frame. A single spacetime diagram suffices to determine what *all* observers see. It is nevertheless instructive to draw separate spacetime diagrams for each observer, making sure that they all agree.

In this chapter, we discuss the two most famous paradoxes in special relativity, the pole and barn paradox and the twin paradox. We also briefly discuss the more subtle aspects of flying manhole covers.

8.2 The Pole and Barn Paradox

A 20-foot pole is moving toward a 10-foot barn fast enough that the pole appears to be only 10 feet long. As soon as both ends of the pole are in the barn, the doors are slammed shut. How can a 20-foot pole fit into a 10-foot barn?

This is the beginning of the pole and barn paradox. It's bad enough to try to imagine what happens to the pole when it suddenly stops and finds itself in a barn that is too small. But what does the pole see?

8. PARADOXES

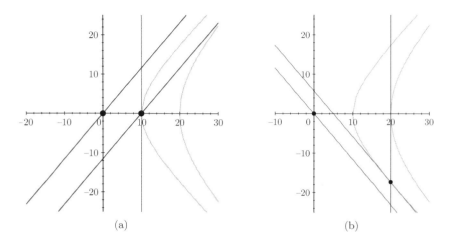

FIGURE 8.1. The pole and barn paradox, (a) in the barn's reference frame and (b) in the pole's reference frame. In each diagram, the heavy straight lines represent the ends of the pole and the lighter straight lines represent the front and back of the barn. The hyperbolas of "radius" 10 and 20 are also shown. The dot at the origin labels the event where the *back* of the pole *enters* the barn, and the other dot labels the event where the *front* of the pole *leaves* the barn.

Length contraction is symmetric, so if the barn sees the pole shortened by a factor of 2, then the pole sees the barn shortened by the same factor of 2. This factor is just $\cosh\beta$, where $|\beta|$ is the same in both cases. So the 20-foot pole sees this 5-foot barn approaching. No way is the pole going to fit in the barn!

The spacetime diagrams for this situation are shown in Figure 8.1(a)–(b), as seen in the barn's reference frame and in the pole's reference frame, respectively. The two dots represent the events of closing the barn doors when the ends of the pole are even with the corresponding door.

In the *barn's* frame (Figure 8.1(a)), these events happen simultaneously, so it is possible in principle to trap the pole in the barn by shutting the doors "at the same time." (We omit speculation about what happens when the pole hits the closed door!)

In the *pole's* frame (Figure 8.1(b)), the exit door is closed long before the rear of the pole enters the barn. Assuming the pole keeps going, for instance by virtue of the door opening again, then the entrance door is closed much later, when the rear of the pole finally gets there. The pole "thinks" it is silly to try to catch it by waiting to close the entrance door until most of the pole has already escaped through the exit door!

8.3 THE TWIN PARADOX

One twin travels 24 lightyears to star X at speed $\frac{24}{25}c$; her twin brother stays home on Earth. When the traveling twin gets to star X, she immediately turns around, and returns at the same speed. How long does each twin think the trip took?

Star X is 24 lightyears away, so, according to the twin at home, it takes her 25 years to get there, and 25 more to return, for a total of 50 years away from Earth. But the traveling twin's clock runs slow by a factor of

$$\cosh\beta = \sqrt{\frac{1}{1-\tanh^2\beta}} = \frac{25}{7}.$$

This means that, according to the traveling twin, it takes her only 7 years each way. Thus, she has aged only 14 years while her brother has aged 50.

This is, in fact, correct, and represents a sort of time travel into the future: it takes the traveling twin 14 years to get 50 years into Earth's future. (Unfortunately, there's no way to get back.)

But wait a minute. The traveling twin should see her brother's clock run slow by the same factor of $\frac{25}{7}$. So when 7 years of her time elapse, she thinks her brother has aged only $\frac{49}{25} \approx 2$ years. Her brother should therefore have aged only 4 years when she returns, from her perspective.

This cannot be right. Either her brother is 4 years older or he is 50 years older. Both siblings must surely agree on that!

The easiest way to resolve this paradox is to draw a single spacetime diagram showing the entire trip in the reference frame of the stay-at-home twin, as shown in Figure 8.2. The lower half of the figure is a hyperbolic triangle with $\tanh\beta = \frac{24}{25}$, and hypotenuse 7 years. The remaining diagonal lines are lines of constant time for the traveling twin, at the point of turnaround, while going and returning, respectively. There is another right triangle containing the hyperbolic angle β; the right angle is at the point of turnaround, and the hypotenuse is $\frac{7}{\cosh\beta} = \frac{49}{25}$ years, the age of the stay-at-home twin "when" the traveling twin turns around.

How much does each twin age? Simply measure the length of their world lines. Intervals are invariant, so it doesn't matter how you compute them. The clear answer is that the brother has indeed aged 50 years, while the sister has aged only 14 years.

So what was wrong with the argument that the brother should have aged only 4 years? There are really three reference frames here: Earth, going, and returning. The "going" and "returning" frames yield different times on Earth for the turnaround—and these times differ by precisely

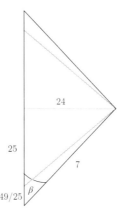

FIGURE 8.2. Spacetime diagram for the twin paradox. The vertical line represents the world line of the stay-at-home twin, and the heavy diagonal lines show the world line of the traveling twin.

$\frac{1152}{25} \approx 46$ years. From this point of view, it takes the traveling twin 46 years to turn around.

One difference between the twins is that the traveling twin is *not* in an inertial frame—she is in two inertial frames, but must accelerate in order to switch from one to the other. This breaks the symmetry between the two twins.

However, this is not really the best way to explain this paradox. It is possible to remove this particular asymmetry by assuming the universe is closed, so that the traveling twin doesn't need to turn around. A simplified version of this is to put the problem on a cylinder [Dray 90]. It turns out that this introduces another sort of asymmetry, but there is a simpler way to look at it.

The amount an observer ages is just the timelike interval measured along his or her worldline. We used this argument here. This approach also works for curved worldlines, corresponding to noninertial observers—except that one must integrate the infinitesimal timelike interval $d\tau = \sqrt{dt^2 - ds^2}$ along the worldline.

A little thought leads to the following remarkable result: *The timelike line connecting two events (assuming there is one) is the* longest *path from one to the other.* (Think about it. Use this line as the t axis. Then any other path has a nonzero contribution from the change in x—which *decreases* the hyperbolic length of the path, and hence the time taken.)

8.4 Manhole Covers

There are two well-known paradoxes involving manhole covers that illustrate some unexpected implications of special relativity. In the first, a 2-foot manhole cover approaches a 2-foot manhole at relativistic velocity. Since the hole sees the cover as much smaller than two feet long, the cover must fall into the manhole. It does.

But what does the cover see? It sees a very small hole rushing at it. No way is this enormous manhole cover going to fit into this small hole.

The resolution of this paradox requires careful consideration of what it means for something to begin to fall, and is left to the reader.

There is also a higher-dimensional version of this problem, without the complication of falling—that is, without gravity. Suppose the manhole cover is moving horizontally as before, but now the hole is in a metal sheet that is rising up to meet it. Again, from the point of view of the hole, the cover is very small and so—if the timing is right—the cover will pass through the hole. It does.

But what does the cover see? It again sees a very small hole rushing at it. How do you get a big object through a small hole? This time we must consider what it means for the cover to "pass through" the hole.

Anyone who successfully resolves these two paradoxes will realize that some properties of materials that we take for granted are quite impossible in special relativity.

➤ CHAPTER 9 ≺

RELATIVISTIC MECHANICS

In which it is shown that mass is energy.

9.1 PROPER TIME

Let τ denote time as measured by a clock carried by an observer moving at constant speed u with respect to the given frame. We call τ "wristwatch time" [Taylor and Wheeler 92], or, more formally, *proper time*. Such a clock simply measures the "length" of the observer's worldline. But this length can be measured in any reference frame.

The (infinitesimal) geometry of this situation is shown in Figure 9.1. Since $d\tau$ is the length of the hypotenuse, it follows immediately that

$$d\tau = \frac{1}{\cosh\alpha}\, dt. \tag{9.1}$$

Equivalently, since the position of the moving observer doesn't change in the moving observer's own rest frame, we have, from the invariance of the interval, that

$$dx^2 - c^2 dt^2 = 0 - c^2 d\tau^2, \tag{9.2}$$

so that

$$d\tau^2 = \left(1 - \frac{1}{c^2}\left(\frac{dx}{dt}\right)^2\right) dt^2, \tag{9.3}$$

or, equivalently,

$$d\tau = \sqrt{1 - \frac{u^2}{c^2}}\, dt = \frac{1}{\gamma}\, dt = \frac{1}{\cosh\alpha}\, dt. \tag{9.4}$$

This computation yields the same answer (for τ, not for t) in any reference frame; proper time is independent of reference frame.

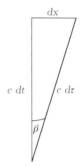

FIGURE 9.1. The geometry of proper time.

9.2 VELOCITY

Consider the *ordinary velocity* of a moving object, defined by

$$u = \frac{d}{dt}x. \tag{9.5}$$

This transforms in a complicated way, since

$$\frac{1}{c}\frac{dx'}{dt'} = \frac{\frac{1}{c}\frac{dx}{dt} - \frac{v}{c}}{1 - \frac{v}{c^2}\frac{dx}{dt}}. \tag{9.6}$$

The reason for this is that both the numerator and the denominator need to be transformed. Note that Equation (9.6) is just the Einstein addition formula for velocities, which we have therefore independently derived by using Lorentz transformations. But that's not quite what we are looking for here.

The invariance of proper time suggests that we should instead differentiate with respect to proper time, since

$$\frac{d}{d\tau}x' = \frac{dx'}{d\tau}. \tag{9.7}$$

In other words, the operator $\frac{d}{d\tau}$ pulls through the Lorentz transformation; only the numerator is transformed when changing reference frames.

Furthermore, the same argument can be applied to t, which suggests that there are (in two dimensions) two components to the velocity. We therefore consider the "2-velocity"

$$\boldsymbol{u} = \frac{d}{d\tau}\begin{pmatrix} ct \\ x \end{pmatrix} = \begin{pmatrix} c\frac{dt}{d\tau} \\ \frac{dx}{d\tau} \end{pmatrix}. \tag{9.8}$$

But since
$$dt = \cosh \alpha \, d\tau \tag{9.9}$$
and
$$dx^2 - c^2 dt^2 = -c^2 d\tau^2, \tag{9.10}$$
we also have
$$dx = c \sinh \alpha \, d\tau, \tag{9.11}$$
so that
$$\boldsymbol{u} = c \begin{pmatrix} \cosh \alpha \\ \sinh \alpha \end{pmatrix}. \tag{9.12}$$

Note that $\frac{1}{c} \boldsymbol{u}$ is a *unit* vector; that is,
$$\frac{1}{c^2} \boldsymbol{u} \cdot \boldsymbol{u} = -1. \tag{9.13}$$

Further,
$$\frac{u}{c} = \frac{dx}{c\,dt} = \tanh \alpha, \tag{9.14}$$
as expected.

9.3 Conservation Laws

Suppose that Newtonian momentum is conserved in a given frame; that is,
$$\sum m_i v_i = \sum M_j V_j. \tag{9.15}$$
(Both of these sums would be zero in the center-of-mass frame.) Changing to another frame moving with respect to the first at speed v, we have
$$v_i = v'_i + v, \tag{9.16}$$
$$V_j = V'_j + v, \tag{9.17}$$
so that
$$\sum m_i(v'_i + v) = \sum M_j(V'_j + v). \tag{9.18}$$
We therefore see that
$$\sum m_i v'_i = \sum M_j V'_j \iff \sum m_i = \sum M_j. \tag{9.19}$$
That is, momentum is conserved in *all* inertial frames provided it is conserved in one frame *and* mass is conserved.

Repeating the computation for kinetic energy, we obtain, starting from

$$\frac{1}{2} \sum m_i v_i^2 = \frac{1}{2} \sum M_j V_j^2, \tag{9.20}$$

that

$$\frac{1}{2} \sum m_i (v_i' + v)^2 = \frac{1}{2} \sum M_j (V_j' + v)^2. \tag{9.21}$$

Expanding this out, we discover that (kinetic) energy is conserved in all frames provided it is conserved in one frame *and* both mass and momentum are conserved.

The situation in special relativity is quite different.

Consider first the momentum defined by the ordinary velocity, namely,

$$p = mu = m\frac{dx}{dt}. \tag{9.22}$$

This momentum is *not* conserved.

We use instead the momentum defined by the 2-velocity, which is given by

$$p = m\frac{dx}{d\tau} = mc \sinh \alpha. \tag{9.23}$$

Suppose now that, as seen in a particular inertial frame, the total momentum of a collection of particles is the same before and after some interaction; that is,

$$\sum m_i c \sinh \alpha_i = \sum M_j c \sinh A_j. \tag{9.24}$$

Consider now the same situation as seen by another inertial reference frame, moving with respect to the first with speed

$$v = c \tanh \beta. \tag{9.25}$$

We therefore have

$$\alpha_i = \alpha_i' + \beta, \tag{9.26}$$
$$A_j = A_j' + \beta. \tag{9.27}$$

Inserting Equation (9.26) into the conservation rule (9.24) leads to

$$\sum m_i c \sinh \alpha_i' = \sum m_i c \sinh(\alpha_i - \beta) \tag{9.28}$$
$$= \left(\sum m_i c \sinh \alpha_i\right) \cosh \beta - \left(\sum m_i c \cosh \alpha_i\right) \sinh \beta,$$

9.4. ENERGY

and similarly inserting Equation (9.27) into (9.24) leads to

$$\sum M_j c \sinh A'_j = \left(\sum M_j c \sinh A_j\right) \cosh \beta - \left(\sum M_j c \cosh A_j\right) \sinh \beta. \tag{9.29}$$

The coefficients of $\cosh \beta$ in these two expressions are equal due to the assumed conservation of momentum in the original frame. We therefore see that conservation of momentum will hold in the new frame if and only if, in addition, the coefficients of $\sinh \beta$ agree, namely,

$$\sum m_i c \cosh \alpha_i = \sum M_j c \cosh A_j. \tag{9.30}$$

But what is this?

9.4 ENERGY

This mystery is resolved by recalling that momentum is mass times velocity, and that there is also a "t-component" to the velocity. In analogy with 2-velocity, we therefore define "2-momentum" to be

$$\boldsymbol{p} = m \begin{pmatrix} c\frac{dt}{d\tau} \\ \frac{dx}{d\tau} \end{pmatrix} = mc \begin{pmatrix} \cosh \alpha \\ \sinh \alpha \end{pmatrix}. \tag{9.31}$$

The second term is clearly the momentum, which we denote by p, but what is the first term? If the object is at rest, $\alpha = 0$, and the first term is therefore just mc. But Einstein's famous equation,

$$E = mc^2, \tag{9.32}$$

leads us to suspect that this is some sort of energy. In fact, mc^2 is called the *rest energy* or *rest mass*.

In general, we *define* the energy of an object moving at speed $u = c \tanh \alpha$ to be the first component of $\boldsymbol{p}$; that is, we define

$$E := mc^2 \cosh \alpha, \tag{9.33}$$
$$p := mc \sinh \alpha, \tag{9.34}$$

or, equivalently,

$$\boldsymbol{p} = \begin{pmatrix} \frac{1}{c}E \\ p \end{pmatrix}. \tag{9.35}$$

FIGURE 9.2. The geometry of 2-momentum.

Is this definition reasonable? Consider the case $\frac{u}{c} \ll 1$. Then

$$E = mc^2 \cosh\alpha = mc^2 \gamma$$
$$= \frac{mc^2}{\sqrt{1 - \frac{u^2}{c^2}}} \qquad (9.36)$$
$$\approx mc^2 + \frac{1}{2}mu^2 + \frac{3}{8}m\frac{u^4}{c^2} + \dots .$$

The first term is the rest energy, the next term is the Newtonian kinetic energy, and the remaining terms are relativistic corrections to the kinetic energy.

The moral is that conservation of 2-momentum is equivalent to both conservation of momentum and conservation of energy, but there is no requirement that the total mass be conserved.

Taking the (squared) norm of the 2-momentum, we obtain

$$-c^2 \boldsymbol{p} \cdot \boldsymbol{p} = E^2 - p^2 c^2 = m^2 c^4. \qquad (9.37)$$

The geometry of this situation is shown in Figure 9.2.

Note that Equation (9.37) continues to makes sense if $m = 0$, even though the expressions for E and p separately in terms of α or γ do not. In fact, γ must approach ∞, or, equivalently, $\frac{u^2}{c^2} = 1$, so that $|u| = c$; such particles *always* move at the speed of light.

We therefore conclude that there can be massless particles that move at the speed of light and satisfy $m = 0$ and

$$E = |p|c \neq 0. \qquad (9.38)$$

Photons are examples of such particles; quantum mechanically, one has $E = \hbar\nu$, where ν is the frequency of the light, and $\hbar = \frac{h}{2\pi}$, where h is Planck's constant.

9.5 USEFUL FORMULAS

The key formulas for analyzing the collision of relativistic particles can all be derived from definitions (9.33) and (9.34).

Taking the difference of squares leads to the key formula (9.37) relating energy, momentum, and (rest) mass, which holds also for massless particles. Rewriting definitions (9.33) and (9.34) leads directly to

$$\gamma = \cosh\alpha = \frac{E}{mc^2} = \frac{1}{\sqrt{1 - \frac{u^2}{c^2}}} \tag{9.39}$$

and

$$\sinh\alpha = \frac{p}{mc} = \frac{u}{c}\gamma, \tag{9.40}$$

and dividing formula (9.40) by (9.39) yields

$$\tanh\alpha = \frac{pc}{E} = \frac{u}{c}. \tag{9.41}$$

≻ Chapter 10 ≺

Problems II

10.1 Mass Isn't Conserved

Problem 10.1. *Two identical lumps of clay of (rest) mass m collide head on, with each moving at $\frac{3}{5}c$. What is the mass of the resulting lump of clay?*

First solution: We assume this is an elastic collision; that is, we do not worry about the details of the actual collision. Conservation of momentum doesn't help here—there is no momentum either before or after the collision. So we need to use conservation of energy. After the collision, there is no kinetic energy, so we have

$$E' = Mc^2. \tag{10.1}$$

Before the collision, we know that the energy of each lump is

$$E = mc^2 \cosh\alpha, \tag{10.2}$$

but how do we find α? We are given that each lump is moving at $\frac{3}{5}c$, so this means we know

$$\tanh\alpha = \frac{3}{5}. \tag{10.3}$$

Yes, we could now use the formula $\cosh\alpha = 1/\sqrt{1-\tanh^2\alpha}$, but it is easier to use a triangle. Since $\tanh\alpha = \frac{3}{5}$, we can scale things so that the legs have "lengths" 3 and 5. Using the hyperbolic Pythagorean theorem, the hypotenuse has length $\sqrt{5^2 - 3^2} = 4$. This is just the triangle in Figure 4.3. Thus,

$$\cosh\alpha = \frac{5}{4} \tag{10.4}$$

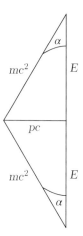

FIGURE 10.1. The momentum diagram for two colliding lumps of clay.

so that
$$Mc^2 = E' = 2E = 2mc^2 \cosh\alpha = \frac{5}{2}mc^2, \qquad (10.5)$$
and therefore
$$M = \frac{5}{2}m. \qquad (10.6)$$

Second solution: Use conservation of 2-momentum, a vector quantity. That is, draw a *momentum diagram*, similar to a spacetime diagram but with units of either momentum (or, more commonly, energy), showing the 2-momenta both before and after the collision. Figure 10.1 shows such a diagram for the two lumps of clay. The initial configuration is represented by the two diagonal lines, one for each lump, moving in opposite directions; the final configuration is represented by the vertical line. Simple triangle trigonometry leads to the same computation as in the first solution, which can also be regarded as vector addition.

10.2 IDENTICAL PARTICLES

PROBLEM 10.2. *Consider the head-on collision of two identical particles, each of mass m and energy E.*

(a) *In Newtonian mechanics, what multiple of E is the energy E' of one particle as observed in the reference frame of the other?*

10.3. Pion Decay I

(b) In special relativity, what is the energy E' of one particle as observed in the reference frame of the other?

(c) Suppose we collide two protons ($mc^2 = 1$ GeV) with energy $E = 30$ GeV. Roughly what multiple of E is E' in this case?

(a) In the center-of-mass frame, each particle has Newtonian kinetic energy
$$E = \frac{1}{2}mv^2. \tag{10.7}$$
In the reference frame of one of the particles, the other particle is moving twice as fast, so that
$$E' = \frac{1}{2}m(2v)^2 = 4E. \tag{10.8}$$

(b) Now we must use the relativistic energy
$$E = mc^2 \cosh \alpha. \tag{10.9}$$
In the reference frame of one of the particles, the other is *not* moving twice as fast. Rather, the hyperbolic angle has doubled. Thus,
$$E' = mc^2 \cosh(2\alpha) = mc^2(2\cosh^2 \alpha - 1), \tag{10.10}$$
so that
$$\frac{E'}{E} = \frac{2\cosh^2 \alpha - 1}{\cosh \alpha}. \tag{10.11}$$
The corresponding momentum diagrams are shown in Figure 10.2 in both frames of reference.

(c) We are given that
$$\cosh \alpha = \frac{E}{mc^2} = 30, \tag{10.12}$$
so that
$$\frac{E'}{E} = \frac{2\cosh^2 \alpha - 1}{\cosh \alpha} \approx 60. \tag{10.13}$$

10.3 Pion Decay I

PROBLEM 10.3. A pion of (rest) mass m and (relativistic) momentum $p = \frac{3}{4}mc$ decays into two (massless) photons. One photon travels in the same direction as the original pion, and the other travels in the opposite direction. Find the energy of each photon.

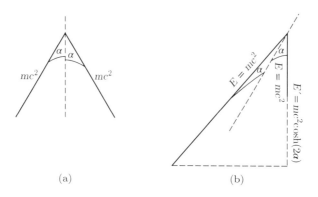

FIGURE 10.2. Momentum diagrams for the collision of identical particles, drawn both (a) in the center-of-mass frame and (b) in the frame of the right-hand particle. In both cases, the dashed line represents the worldline of an observer in the center-of-mass frame.

First solution: We begin by determining the energy and momentum of the pion. Since $p = \frac{3}{4}mc$, we have immediately that

$$\sinh\beta = \frac{p}{mc} = \frac{3}{4}, \qquad (10.14)$$

from which it follows that

$$\cosh\beta = \frac{5}{4} \qquad (10.15)$$

by simple (hyperbolic) triangle trigonometry. Thus, the total energy is given by

$$E = mc^2 \cosh\beta = \frac{5}{4}mc^2. \qquad (10.16)$$

Since each photon satisfies $E_i = |p_i|c$, and since, without loss of generality, we can assume $p_1 > 0$ and $p_2 < 0$, conservation of energy-momentum tells us that

$$E_1 - E_2 = pc = \frac{3}{4}mc^2 \qquad (10.17)$$

and

$$E_1 + E_2 = E = \frac{5}{4}mc^2, \qquad (10.18)$$

from which it follows immediately that

$$E_1 = mc^2 \qquad (10.19)$$

and

$$E_2 = \frac{mc^2}{4}. \qquad (10.20)$$

10.3. Pion Decay I

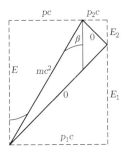

FIGURE 10.3. Momentum diagram for the decay of a pion into photons.

Second solution: Use a momentum diagram, as shown in Figure 10.3. The computation is the same as in the first solution.

Third solution: Use hyperbola geometry and triangle trigonometry, as shown in Figure 10.4 in both the given reference frame (a) and that of the pion (b), taking advantage of the fact that, in the center-of-mass frame (the rest frame of the pion), the two photons must have the same energy E_0 and equal but opposite momenta $\pm p_0 c = \pm E_0$. The details are left to the reader.

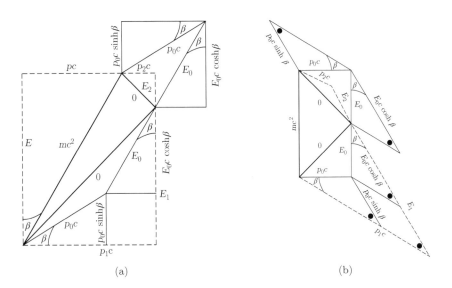

FIGURE 10.4. Analyzing pion decay by using hyperbola geometry, (a) in the laboratory frame and (b) in the rest frame of the pion.

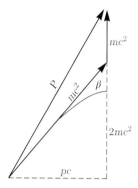

FIGURE 10.5. Spacetime diagram for the collision of identical particles.

10.4 MASS AND ENERGY

PROBLEM 10.4. *A particle of mass m whose total energy is twice its rest energy collides with an identical particle at rest. If they stick together, what is the mass of the resulting composite particle, and what is its speed?*

A momentum diagram for this collision is shown in Figure 10.5. The particle is shown moving to the right; its rest energy mc^2 is given by the magnitude of the hypotenuse of the right triangle shown. The observed energy, however, is $2mc^2$; this is the magnitude of the vertical leg of the triangle. The momentum vector of the second particle is vertical (since it is at rest), with magnitude mc^2 (the rest energy). Adding these two momentum vectors together yields the vector labeled P, whose vertical component is clearly $3mc^2$ and whose horizontal component is the same as the original momentum. We have

$$\cosh\beta = 2, \tag{10.21}$$

and therefore

$$p = mc\sinh\beta = \sqrt{3}\,mc. \tag{10.22}$$

The rest energy of the resulting particle is therefore given by

$$M^2 c^4 = (3mc^2)^2 - (\sqrt{3}mc)^2 c^2 = 6\,m^2 c^4, \tag{10.23}$$

so that $M = \sqrt{6}m$. The speed of the resulting particle is obtained from

$$\frac{v}{c} = \tanh\alpha = \frac{pc}{E} = \frac{\sqrt{3}}{3}. \tag{10.24}$$

10.5 Pion Decay II

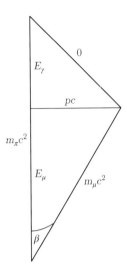

FIGURE 10.6. Spacetime diagram for the decay of a pion.

10.5 Pion Decay II

PROBLEM 10.5. *A pion (mass m_π) at rest decays into a muon (mass m_μ) and a massless neutrino ($m_\nu = 0$). Find the momentum p, the energy E, and the speed v/c of the muon in terms of m_π and m_μ.*

This problem can be solved by brute-force algebra; we present a more geometric solution. A momentum diagram showing this collision is given in Figure 10.6. The vertical line represents the pion at rest; the magnitude of this line is the pion energy $m_\pi c^2$. The muon is shown moving to the right at speed $\frac{v}{c} = \tanh\beta$; the magnitude of the hypotenuse of the lower triangle is the muon energy $m_\mu c^2$. The upper triangle is isosceles, showing the massless neutrino; the vertical leg of this triangle is the neutrino energy $|p|c$.

Conservation of energy-momentum is built into this diagram; we have, first of all, that
$$|p|c = p_\mu c = m_\mu c^2 \sinh\beta, \tag{10.25}$$
and then that
$$m_\pi c^2 = E_\mu + |p|c = m_\mu c^2(\cosh\beta + \sinh\beta) = m_\mu c^2 e^\beta. \tag{10.26}$$
We have therefore shown that
$$e^\beta = \frac{m_\pi}{m_\mu}, \tag{10.27}$$

so that

$$\sinh\beta = \frac{e^\beta - e^{-\beta}}{2} = \frac{m_\pi^2 - m_\mu^2}{2\,m_\mu\,m_\pi}, \tag{10.28}$$

$$\cosh\beta = \frac{e^\beta + e^{-\beta}}{2} = \frac{m_\pi^2 + m_\mu^2}{2\,m_\mu\,m_\pi}, \tag{10.29}$$

$$\tanh\beta = \frac{\sinh\beta}{\cosh\beta} = \frac{m_\pi^2 - m_\mu^2}{m_\pi^2 + m_\mu^2}. \tag{10.30}$$

Thus,

$$p_\mu = m_\mu c \sinh\beta = \frac{m_\pi^2 - m_\mu^2}{2\,m_\pi}\,c, \tag{10.31}$$

$$E_\mu = m_\mu c^2 \cosh\beta = \frac{m_\pi^2 + m_\mu^2}{2\,m_\pi}\,c^2, \tag{10.32}$$

$$\frac{v}{c} = \tanh\beta = \frac{m_\pi^2 - m_\mu^2}{m_\pi^2 + m_\mu^2}. \tag{10.33}$$

Alternatively, we can avoid the use of e^β by rewriting Equation (10.26) in the form

$$m_\pi - m_\mu \cosh\beta = m_\mu \sinh\beta. \tag{10.34}$$

Squaring both sides of this equation and using the fundamental relationship $\cosh^2\beta - \sinh^2\beta = 1$ quickly leads to Equation (10.29), from which $\sinh\beta$ and $\tanh\beta$ can be obtained by using hyperbolic triangle trigonometry.

≻ Chapter 11 ≺

Relativistic Electromagnetism

In which it is shown that electricity and magnetism can no more be separated than space and time.

11.1 Magnetism from Electricity

Our starting point involves the electric and magnetic fields of an infinite straight wire, which are derived in most introductory textbooks on electrodynamics, such as Griffiths [Griffiths 99], and which we state here without proof.

The electric field of an infinite straight wire (in vacuum) with (positive) charge density λ points away from the wire with magnitude

$$E = \frac{\lambda}{2\pi\epsilon_0 r}, \qquad (11.1)$$

where r is the perpendicular distance from the wire and ϵ_0 is the permittivity constant. The magnetic field of such a wire with (positive) current density I has magnitude

$$B = \frac{\mu_0 I}{2\pi r}, \qquad (11.2)$$

where r is again the perpendicular distance from the wire and μ_0 is the permeability constant, which is related to ϵ_0 by

$$\epsilon_0 \mu_0 = \frac{1}{c^2}. \qquad (11.3)$$

(The direction of the magnetic field is obtained as the cross product of the direction of the current and the position vector from the wire to the point in question.)

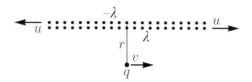

FIGURE 11.1. A wire at rest containing postive charges with charge density λ moving to the right and negative charges with charge density $-\lambda$ moving to the left, each with speed u. A test charge q moving to the right with speed v is also shown.

We also need the *Lorentz force law*, which says that the force $\vec{F}$ on a test particle of charge q and velocity $\vec{v}$ is given by

$$\vec{F} = q(\vec{E} + \vec{v} \times \vec{B}), \tag{11.4}$$

where $\vec{E}$ and $\vec{B}$ denote the electric and magnetic fields with magnitudes E and B, respectively.

Consider an infinite line charge consisting of identical particles of charge $\rho > 0$ separated by a distance ℓ. This gives an infinite wire with (average) charge density

$$\lambda_0 = \frac{\rho}{\ell}. \tag{11.5}$$

Suppose now that the charges are moving to the right with speed

$$u = c \tanh \alpha. \tag{11.6}$$

Due to length contraction, the charge density seen by an observer at rest *increases* to

$$\lambda = \frac{\rho}{\frac{\ell}{\cosh \alpha}} = \lambda_0 \cosh \alpha. \tag{11.7}$$

Suppose now that, as shown in Figure 11.1, there are positively charged particles moving to the right, and equally but negatively charged particles moving to the left, each with speed u. Consider further a test particle of charge $q > 0$ situated a distance r from the wire and moving with speed

$$v = c \tanh \beta \tag{11.8}$$

to the right. Then the net charge density in the laboratory frame is 0, so there is no electrical force on the test particle in this frame. There is, of course, a net current density, namely,

$$I = \lambda u + (-\lambda)(-u) = 2\lambda u. \tag{11.9}$$

11.1. MAGNETISM FROM ELECTRICITY

What does the test particle see? If we switch to the rest frame of the test particle, the negative charges appear to move faster, with speed $u_- > u$, and the positive charges appear to move slower, with speed $u_+ < u$. The relative speeds satisfy

$$\frac{u_+}{c} = \tanh(\alpha - \beta), \tag{11.10}$$

$$\frac{u_-}{c} = \tanh(\alpha + \beta), \tag{11.11}$$

resulting in current densities

$$\lambda_\pm = \lambda \cosh(\alpha \mp \beta) = \lambda(\cosh\alpha\cosh\beta \mp \sinh\alpha\sinh\beta), \tag{11.12}$$

which, in turn, result in a total charge density of

$$\begin{aligned}\lambda' &= \lambda_+ - \lambda_- \\ &= -2\lambda_0 \sinh\alpha \sinh\beta \\ &= -2\lambda \tanh\alpha \sinh\beta.\end{aligned} \tag{11.13}$$

According to Equation (11.1), this then results in an electric field of magnitude

$$E' = \frac{|\lambda'|}{2\pi\epsilon_0 r}, \tag{11.14}$$

which in turn leads to an electric force of magnitude

$$\begin{aligned}F' = qE' &= \frac{\lambda}{\pi\epsilon_0 r} q \tanh\alpha \sinh\beta \\ &= \frac{\lambda u}{\pi\epsilon_0 c^2 r} qv \cosh\beta \\ &= \frac{\mu_0 I}{2\pi r} qv \cosh\beta.\end{aligned} \tag{11.15}$$

To relate this to the force observed in the laboratory frame, we must consider how force transforms under a Lorentz transformation. We have[1]

$$\vec{F}' = \frac{d\vec{p}'}{dt'} \tag{11.16}$$

and, of course,

$$\vec{F} = \frac{d\vec{p}}{dt}. \tag{11.17}$$

---

[1] This is the traditional notion of force, which does not transform simply between frames. A possibly more useful notion of force is obtained by differentiating with respect to proper time, as discussed briefly in Section 11.6.

But since in this case the force is perpendicular to the direction of motion, we have
$$d\vec{p} = d\vec{p}', \tag{11.18}$$
and since $dx' = 0$ in the comoving frame, we also have
$$dt = dt' \cosh\beta. \tag{11.19}$$
Thus, in this case, the magnitudes are related by
$$F = \frac{F'}{\cosh\beta} = -\frac{\mu_0 I}{2\pi r} qv. \tag{11.20}$$
But this is just the Lorentz force law
$$\vec{F} = q\,\vec{v} \times \vec{B} \tag{11.21}$$
with $B = |\vec{B}|$ given by Equation (11.2).

We conclude that in the laboratory frame there is a *magnetic* force on the test particle, which is just the *electric* force observed in the comoving frame.

11.2 LORENTZ TRANSFORMATIONS

We now investigate more general transformations of electric and magnetic fields between different inertial frames. Our starting point is the electromagnetic field of an infinite flat metal sheet, which is derived in most introductory textbooks on electrodynamics, such as Griffiths [Griffiths 99], and which we state here without proof.

The electric field of an infinite metal sheet with (positive) charge density σ points away from the sheet and has constant magnitude
$$E = \frac{\sigma}{2\epsilon_0}. \tag{11.22}$$

The magnetic field of such a sheet with current density $\vec{\kappa}$ has constant magnitude
$$B = \frac{\mu_0}{2}|\vec{\kappa}| \tag{11.23}$$
and direction determined by the right-hand rule.

Consider a capacitor consisting of two horizontal ($y = $ constant) parallel plates, with equal and opposite charge densities, as shown in Figure 11.2. For definiteness, take the charge density on the bottom plate to be $\sigma_0 > 0$,

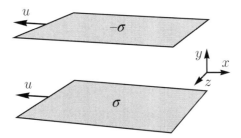

FIGURE 11.2. A capacitor moving to the left with speed u. The bottom plate has charge density $\sigma > 0$, and the top plate has charge density $-\sigma$.

and suppose that the charges are at rest—that is, suppose that the current density of each plate is zero. Then the electric field is given by

$$\vec{E}_0 = E_0\,\hat{y} = \frac{\sigma_0}{\epsilon_0}\,\hat{y} \tag{11.24}$$

between the plates, and it vanishes elsewhere. Now let the capacitor move to the left with velocity

$$\vec{u} = -u\,\hat{x} = -c\tanh\alpha\,\hat{x}. \tag{11.25}$$

Then the *width* of the plate is unchanged, but, just as for the line charge in Equation (11.7), the *length* is Lorentz contracted, which *decreases* the area and hence *increases* the charge density. The charge density (on the bottom plate) is therefore

$$\sigma = \sigma_0 \cosh\alpha. \tag{11.26}$$

But there is now also a current density, which is given by

$$\vec{\kappa} = \sigma\,\vec{u} \tag{11.27}$$

on the lower plate. The top plate has charge density $-\sigma$, so its current density is $-\vec{\kappa}$. Then both the electric and magnetic fields vanish outside the plates, whereas inside the plates we have

$$\vec{E} = E^y\,\hat{y} = \frac{\sigma}{\epsilon_0}\,\hat{y}, \tag{11.28}$$

$$\vec{B} = B^z\,\hat{z} = -\mu_0\,\sigma u\,\hat{z}, \tag{11.29}$$

which can be rewritten by using Equations (11.25) and (11.26), in the form

$$E^y = E_0 \cosh \alpha, \quad (11.30)$$
$$B^z = B_0 \sinh \alpha. \quad (11.31)$$

For later convenience, we have introduced in the last equation the quantity

$$B_0 = -c\mu_0 \sigma_0 = -c\mu_0 \epsilon_0 E_0 = -\frac{1}{c} E_0, \quad (11.32)$$

which does *not* correspond to the magnetic field when the plate is at rest—which, of course, vanishes since $\vec{u} = 0$.

The preceding discussion gives the electric and magnetic fields seen by an observer at rest. But what is seen by an observer moving to the right with speed $v = c \tanh \beta$? To compute this, we first use the velocity addition law to compute the correct rapidity to insert in Equations (11.30) and (11.31), which is simply the sum of the rapidities α and β.

The moving observer therefore sees an electric field $\vec{E}'$ and a magnetic field $\vec{B}'$. From Equations (11.30)–(11.32) and the hyperbolic trigonometric formulas (4.5) and (4.6), we have

$$\begin{aligned} E'^y &= E_0 \cosh(\alpha + \beta) \\ &= E_0 \cosh \alpha \cosh \beta + E_0 \sinh \alpha \sinh \beta \\ &= E_0 \cosh \alpha \cosh \beta - cB_0 \sinh \alpha \sinh \beta \\ &= E^y \cosh \beta - cB^z \sinh \beta, \end{aligned} \quad (11.33)$$

and similarly,

$$\begin{aligned} B'^z &= B_0 \sinh(\alpha + \beta) \\ &= B_0 \sinh \alpha \cosh \beta + B_0 \cosh \alpha \sinh \beta \\ &= B^z \cosh \beta - \frac{1}{c} E^y \sinh \beta. \end{aligned} \quad (11.34)$$

The last step in each case is crucial; we eliminate α, E_0, and B_0, since the "lab" frame is an artifact of our construction. The resulting transformations depend, as desired, on only the two "moving" frames and their relative rapidity, β.

Repeating the argument with the y- and z-axes interchanged (and being careful about the orientation), we obtain the analogous formulas

$$E'^z = E^z \cosh \beta + cB^y \sinh \beta, \quad (11.35)$$
$$B'^y = B^y \cosh \beta + \frac{1}{c} E^z \sinh \beta. \quad (11.36)$$

11.3. VECTORS

Finally, by considering motion perpendicular to the plates, we can show [Griffiths 99]

$$E'^x = E^x, \qquad (11.37)$$

and by considering a solenoid, we obtain [Griffiths 99]

$$B'^x = B^x. \qquad (11.38)$$

Equations (11.33)–(11.38) describe the behavior of the electric and magnetic fields under Lorentz transformations. These equations can be nicely rewritten in vector language by introducing the projections parallel and perpendicular to the direction of motion of the observer, namely,

$$\vec{E}_\| = \frac{\vec{v} \cdot \vec{E}}{\vec{v} \cdot \vec{v}} \vec{v}, \qquad (11.39)$$

$$\vec{B}_\| = \frac{\vec{v} \cdot \vec{B}}{\vec{v} \cdot \vec{v}} \vec{v}, \qquad (11.40)$$

and

$$\vec{E}_\perp = \vec{E} - \vec{E}_\|, \qquad (11.41)$$

$$\vec{B}_\perp = \vec{B} - \vec{B}_\|. \qquad (11.42)$$

We then have

$$\vec{E}'_\| = \vec{E}_\|, \qquad (11.43)$$

$$\vec{B}'_\| = \vec{B}_\|, \qquad (11.44)$$

and

$$\vec{B}'_\perp = \left(\vec{B}_\perp - \frac{1}{c^2} \vec{v} \times \vec{E}_\perp \right) \cosh\beta, \qquad (11.45)$$

$$\vec{E}'_\perp = \left(\vec{E}_\perp + \vec{v} \times \vec{B}_\perp \right) \cosh\beta. \qquad (11.46)$$

11.3 VECTORS

In Chapters 5 and 9, we used 2-component vectors to describe spacetime quantities such as position and velocity, with one component for time and the other for space. In the case of three spatial dimensions, we use

4-component vectors, namely,

$$x^\nu = \begin{pmatrix} x^0 \\ x^1 \\ x^2 \\ x^3 \end{pmatrix} = \begin{pmatrix} ct \\ x \\ y \\ z \end{pmatrix}. \tag{11.47}$$

These are called *contravariant* vectors, and their indices are written "upstairs," that is, as superscripts.

Just as before, Lorentz transformations are hyperbolic rotations, which must now be written as 4×4 matrices. For instance, a "boost" in the x direction now takes the form

$$\begin{pmatrix} ct' \\ x' \\ y' \\ z' \end{pmatrix} = \begin{pmatrix} \cosh\beta & -\sinh\beta & 0 & 0 \\ -\sinh\beta & \cosh\beta & 0 & 0 \\ 0 & 0 & 1 & 0 \\ 0 & 0 & 0 & 1 \end{pmatrix} \begin{pmatrix} ct \\ x \\ y \\ z \end{pmatrix}. \tag{11.48}$$

A general Lorentz transformation can be written in the form

$$x'^\mu = \Lambda^\mu{}_\nu x^\nu, \tag{11.49}$$

where the $\Lambda^\mu{}_\nu$ are (the components of) the appropriate 4×4 matrix, and where we have adopted the *Einstein summation convention* that repeated indices, in this case ν, are to be summed from 0 to 3. In matrix notation, this can be written as

$$\boldsymbol{x}' = \boldsymbol{\Lambda}\boldsymbol{x}. \tag{11.50}$$

Why are some indices "up" and others "down"? In relativity, both special and general, it is essential to distinguish between two types of vectors. In addition to contravariant vectors, there are also *covariant* vectors, often referred to as dual vectors. The dual vector associated with x^μ is[2]

$$x_\mu = \begin{pmatrix} -x_0 & x_1 & x_2 & x_3 \end{pmatrix} = \begin{pmatrix} -ct & x & y & z \end{pmatrix}. \tag{11.51}$$

We won't have much need for covariant vectors, but note that the invariance of the interval can be nicely written as

$$\begin{aligned} x_\mu x^\mu &= -c^2 t^2 + x^2 + y^2 + z^2 \\ &= x'_\mu x'^\mu. \end{aligned} \tag{11.52}$$

(Don't forget the summation convention.) In fact, this property can be taken as the *definition* of Lorentz transformations, and it is straightforward to determine which matrices $\Lambda^\mu{}_\nu$ are allowed.

---

[2]Some authors use different conventions.

Taking the derivative with respect to proper time leads to the *4-velocity*

$$u^\mu = \frac{dx^\mu}{d\tau} = \frac{dx^\mu}{dt}\frac{dt}{d\tau}. \tag{11.53}$$

It is often useful to divide the components of the 4-velocity into space and time in the form

$$u = \begin{pmatrix} c\gamma \\ \vec{v}\gamma \end{pmatrix} = \begin{pmatrix} c\cosh\beta \\ \hat{v}\,c\sinh\beta \end{pmatrix}, \tag{11.54}$$

where $\hat{v}$ is the unit vector in the direction of $\vec{v}$. Note that the 4-velocity is a unit vector in the sense that

$$\frac{1}{c^2}u_\mu u^\mu = -1. \tag{11.55}$$

The 4-momentum is is simply the 4-velocity times the rest mass; that is,

$$p^\mu = mu^\mu = \begin{pmatrix} \frac{1}{c}E \\ \vec{p} \end{pmatrix} = \begin{pmatrix} mc\gamma \\ m\vec{v}\gamma \end{pmatrix} = \begin{pmatrix} mc\cosh\beta \\ \hat{v}\,mc\sinh\beta \end{pmatrix}. \tag{11.56}$$

Note that

$$p_\mu p^\mu = -m^2 c^2, \tag{11.57}$$

which is equivalent to our earlier result

$$E^2 - p^2 c^2 = m^2 c^4. \tag{11.58}$$

11.4 TENSORS

Roughly speaking, tensors are like vectors, but with more components and hence more indices. We consider here only one particular case, namely, *rank-2 contravariant tensors*, which have two "upstairs" indices. In a particular reference frame, the components of such a tensor make up a 4 × 4 matrix,

$$T^{\mu\nu} = \begin{pmatrix} T^{00} & T^{01} & T^{02} & T^{03} \\ T^{10} & T^{11} & T^{12} & T^{13} \\ T^{20} & T^{21} & T^{22} & T^{23} \\ T^{30} & T^{31} & T^{32} & T^{33} \end{pmatrix}. \tag{11.59}$$

How does the tensor **T** transform under Lorentz transformations? Well, it has *two* indices, *each* of which must be transformed. This leads to a transformation of the form

$$T'^{\mu\nu} = \Lambda^\mu{}_\rho \Lambda^\nu{}_\sigma T^{\rho\sigma} = \Lambda^\mu{}_\rho T^{\rho\sigma} \Lambda^\nu{}_\sigma, \tag{11.60}$$

where the second form (and the summation convention) leads naturally to the matrix equation
$$\boldsymbol{T}' = \boldsymbol{\Lambda T \Lambda}^t, \tag{11.61}$$
where t denotes matrix transpose.

Further simplification occurs in the special case where $\boldsymbol{T}$ is antisymmetric; that is,
$$T^{\nu\mu} = -T^{\mu\nu}, \tag{11.62}$$
so that the components of $\boldsymbol{T}$ take the form
$$T^{\mu\nu} = \begin{pmatrix} 0 & a & b & c \\ -a & 0 & f & -e \\ -b & -f & 0 & d \\ -c & e & -d & 0 \end{pmatrix}. \tag{11.63}$$

11.5 THE ELECTROMAGNETIC FIELD

Why have we done all this? Well, first of all, note that, due to antisymmetry, $\boldsymbol{T}$ has precisely six independent components. Next, compute $\boldsymbol{T}'$ by using matrix multiplication and the fundamental hyperbolic trigonometric identity (4.4). As you should check for yourself, the result is
$$T'^{\mu\nu} = \begin{pmatrix} 0 & a' & b' & c' \\ -a' & 0 & f' & -e' \\ -b' & -f' & 0 & d' \\ -c' & e' & -d' & 0 \end{pmatrix}, \tag{11.64}$$
where
$$a' = a, \tag{11.65}$$
$$b' = b\cosh\beta - f\sinh\beta, \tag{11.66}$$
$$c' = c\cosh\beta + e\sinh\beta, \tag{11.67}$$
$$d' = d, \tag{11.68}$$
$$e' = e\cosh\beta + c\sinh\beta, \tag{11.69}$$
$$f' = f\cosh\beta - b\sinh\beta. \tag{11.70}$$

Equations (11.65)–(11.67) are similar to the transformation rule for the electric field, and Equations (11.68)–(11.70) are similar to the transformation rule for the magnetic field!

We conclude that the electromagnetic field is described by an antisymmetric, rank-2 tensor of the form

$$F^{uv} = \begin{pmatrix} 0 & \frac{1}{c}E^x & \frac{1}{c}E^y & \frac{1}{c}E^z \\ -\frac{1}{c}E^x & 0 & B^z & -B^y \\ -\frac{1}{c}E^y & -B^z & 0 & B^x \\ -\frac{1}{c}E^z & B^y & -B^x & 0 \end{pmatrix}, \qquad (11.71)$$

which is known as the *electromagnetic field tensor*.

11.6 Maxwell's Equations

Maxwell's equations in vacuum (and in MKS units) are

$$\vec{\nabla} \cdot \vec{E} = \frac{1}{\epsilon_0} \rho, \qquad (11.72)$$

$$\vec{\nabla} \cdot \vec{B} = 0, \qquad (11.73)$$

$$\vec{\nabla} \times \vec{E} = -\frac{\partial \vec{B}}{\partial t}, \qquad (11.74)$$

$$\vec{\nabla} \times \vec{B} = \mu_0 \vec{J} + \mu_0 \epsilon_0 \frac{\partial \vec{E}}{\partial t}, \qquad (11.75)$$

where ρ is the charge density, $\vec{J}$ is the current density, and the constants μ_0 and ϵ_0 satisfy Equation (11.3). Equation (11.72) is just the differential form of Gauss's law, (11.74) is Faraday's equation, and (11.75) is Ampère's law corrected for the case of a time-dependent electric field. We also have the charge conservation equation

$$\vec{\nabla} \cdot \vec{J} = -\frac{\partial \rho}{\partial t} \qquad (11.76)$$

and the Lorentz force law

$$\vec{F} = q(\vec{E} + \vec{v} \times \vec{B}). \qquad (11.77)$$

Maxwell's equations (11.73) and (11.74) are automatically solved by introducing the scalar potential Φ and the vector potential $\vec{A}$ and defining

$$\vec{B} = \vec{\nabla} \times \vec{A}, \qquad (11.78)$$

$$\vec{E} = -\frac{\partial \vec{A}}{\partial t} - \vec{\nabla}\Phi. \qquad (11.79)$$

What form do Maxwell's equations take if we rewrite them in tensor language? Consider the following derivatives of $F^{\mu\nu}$:

$$\frac{\partial F^{\mu\nu}}{\partial x^\nu} = \frac{\partial F^{\mu 0}}{\partial t} + \frac{\partial F^{\mu 1}}{\partial x} + \frac{\partial F^{\mu 2}}{\partial y} + \frac{\partial F^{\mu 3}}{\partial z}, \quad (11.80)$$

which corresponds to four different expressions, one for each value of μ. For $\mu = 0$, we get

$$0 + \frac{1}{c}\frac{\partial E^x}{\partial x} + \frac{1}{c}\frac{\partial E^y}{\partial y} + \frac{1}{c}\frac{\partial E^z}{\partial z} = \frac{1}{c}\vec{\nabla}\cdot\vec{E} = \frac{\rho}{c\epsilon_0} = c\mu_0\rho, \quad (11.81)$$

where Gauss's law was used to get the next-to-last equality. Similarly, for $\mu = 1$, we have

$$-\frac{1}{c^2}\frac{\partial E^x}{\partial t} + 0 + \frac{\partial B^z}{\partial y} - \frac{\partial B^y}{\partial z}. \quad (11.82)$$

Combining Equation (11.82) with the expressions for $\mu = 2$ and $\mu = 3$ yields the left-hand side of

$$-\frac{1}{c^2}\frac{\partial \vec{E}}{\partial t} + \nabla \times \vec{B} = \mu_0 \vec{J}, \quad (11.83)$$

where the right-hand side follows from Ampère's law. Combining these equations, and introducing the *4-current density*

$$J^\mu = \begin{pmatrix} c\rho \\ \vec{J} \end{pmatrix} \quad (11.84)$$

leads to

$$\frac{\partial F^{\mu\nu}}{\partial x^\nu} = \mu_0 J^\mu, \quad (11.85)$$

which is equivalent to the two Maxwell equations with a physical source, namely, Gauss's law (11.72) and Ampère's law (11.75).

Furthermore, taking the four-dimensional divergence of the 4-current density leads to

$$\mu_0 \frac{\partial J^\mu}{\partial x^\mu} = \frac{\partial}{\partial x^\mu}\frac{\partial F^{\mu\nu}}{\partial x^\nu} = 0, \quad (11.86)$$

since there is an implicit double sum over both μ and ν, and the derivatives commute but $F^{\mu\nu}$ is antisymmetric. (Check this by interchanging the order of summation.) Working out the components of this equation, we have

$$\frac{1}{c}\frac{\partial J^0}{\partial t} + \frac{\partial J^1}{\partial x} + \frac{\partial J^2}{\partial y} + \frac{\partial J^3}{\partial z} = 0, \quad (11.87)$$

which is just the charge conservation equation (11.76).

11.6. Maxwell's Equations

What about the remaining equations? Introducing the *dual tensor* $G^{\mu\nu}$ obtained from $F^{\mu\nu}$ (Equation (11.71)) by replacing $\frac{1}{c}\vec{E}$ by $\vec{B}$ and $\vec{B}$ by $-\frac{1}{c}\vec{E}$ results in

$$G^{\mu\nu} = \begin{pmatrix} 0 & B^x & B^y & B^z \\ -B^x & 0 & -\frac{1}{c}E^z & \frac{1}{c}E^y \\ -B^y & \frac{1}{c}E^z & 0 & -\frac{1}{c}E^x \\ -B^z & -\frac{1}{c}E^y & \frac{1}{c}E^x & 0 \end{pmatrix}. \tag{11.88}$$

Then the four equations

$$\frac{\partial G^{\mu\nu}}{\partial x^\nu} = 0 \tag{11.89}$$

correspond to

$$\vec{\nabla} \cdot \vec{B} = 0, \tag{11.90}$$

$$-\frac{1}{c}\frac{\partial \vec{B}}{\partial t} - \frac{1}{c}\vec{\nabla} \times \vec{E} = 0, \tag{11.91}$$

which are precisely the two remaining Maxwell equations, (11.73) and (11.74), respectively.

Thus, Maxwell's equations in tensor form are just (11.85) and (11.89)—two tensor equations rather than four vector equations.

The Lorentzian interval as written in Equation (11.52) can be thought of as the dot product $\boldsymbol{x} \cdot \boldsymbol{x}$ and is a scalar invariant. Two important scalar invariants of the electromagnetic field can be constructed naturally in tensor language, namely,

$$\frac{1}{2}F_{\mu\nu}F^{\mu\nu} = -\frac{1}{c^2}|\vec{E}|^2 + |\vec{B}|^2 = -\frac{1}{2}G_{\mu\nu}G^{\mu\nu} \tag{11.92}$$

and

$$\frac{1}{4}G_{\mu\nu}F^{\mu\nu} = -\frac{1}{c}\vec{E} \cdot \vec{B}, \tag{11.93}$$

where we must take care with the signs of the components of the *covariant* tensors $F_{\mu\nu}$ and $G_{\mu\nu}$. These expressions can be thought of as $\boldsymbol{F} \cdot \boldsymbol{F}$ and $\boldsymbol{F} \cdot \boldsymbol{G}$.

Finally, it is possible to solve the sourcefree Maxwell equations by introducing a *4-potential*

$$A^\mu = \begin{pmatrix} \frac{1}{c}\Phi \\ \vec{A} \end{pmatrix} \tag{11.94}$$

and defining
$$F^{\mu\nu} = \frac{\partial A^\nu}{\partial x_\mu} - \frac{\partial A^\mu}{\partial x_\nu}, \tag{11.95}$$
where again we must take care with the signs of components with "downstairs" indices. Furthermore, the Lorentz force law can be rewritten in the form
$$m\frac{\partial p^\mu}{\partial \tau} = qu_\nu F^{\mu\nu}. \tag{11.96}$$
Note the appearance of the proper time τ in this equation. Just as in Chapter 9, this is because differentiation with respect to τ pulls through a Lorentz transformation, which makes this a valid tensor equation that holds in any inertial frame.

11.7 THE UNIFICATION OF SPECIAL RELATIVITY

In our brief tour of relativistic electromagnetism, we have seen how special relativity unifies physical concepts previously regarded as separate. Space and time become spacetime. Energy and momentum become 4-momentum. Charge and current densities become the 4-current density. The scalar and vector potentials become the 4-potential. And, finally, the electric and magnetic fields become the electromagnetic field tensor.

⪢ Chapter 12 ⪡

Problems III

12.1 Vanishing Fields

PROBLEM 12.1. *Suppose you know that in a particular inertial frame neither the electric field $\vec{E}$ nor the magnetic field $\vec{B}$ has an x component, but neither $\vec{E}$ nor $\vec{B}$ is zero. Consider another inertial frame moving with respect to the first one with velocity v in the x-direction, and denote the electric and magnetic fields in this frame by $\vec{E}'$ and $\vec{B}'$, respectively.*

(a) *What are the conditions on $\vec{E}$ and $\vec{B}$, if any, and the value(s) of v, if any, such that $\vec{E}'$ vanishes for some value of v?*

(b) *What are the conditions on $\vec{E}$ and $\vec{B}$, if any, and the value(s) of v, if any, such that $\vec{B}'$ vanishes for some value of v?*

(c) *What are the conditions on $\vec{E}$ and $\vec{B}$, if any, and the value(s) of v, if any, such that both $\vec{E}'$ and $\vec{B}'$ vanish for the same value of v?*

(d) *Is it possible that $\vec{E}'$ and $\vec{B}''$ vanish for different values of v? (We write $\vec{B}''$ rather than $\vec{B}'$ to emphasize that $\vec{E}'$ and $\vec{B}''$ are with respect to different reference frames.)*

(a) By inserting $\vec{E}' = 0$ into Equation (11.46), we get

$$v = |\vec{v}| = \frac{|\vec{E}|}{|\vec{B}|}, \qquad (12.1)$$

since $\vec{v}$ is perpendicular to $\vec{B}$. This is possible only if $|\vec{E}| < c|\vec{B}|$, which also follows immediately from the invariance of property (11.92).

(b) By inserting $\vec{B}' = 0$ into Equation (11.45), we get

$$v = |\vec{v}| = \frac{c^2|\vec{B}|}{|\vec{E}|}, \tag{12.2}$$

which is possible only if $c|\vec{B}| < |\vec{E}|$. This condition also follows immediately from the invariance of property (11.92).

(c) This is not possible; if the electric and magnetic fields are both zero in any frame, they are zero in all frames.

(d) No. The conditions in parts (a) and (b) cannot both be satisfied.

12.2 Parallel and Perpendicular Fields

PROBLEM 12.2. *Suppose that in a particular inertial frame, the electric field $\vec{E}$ and magnetic field $\vec{B}$ are neither perpendicular nor parallel to each other.*

(a) *Is there another inertial frame in which the fields $\vec{E}''$ and $\vec{B}''$ are parallel to each other?*

(b) *Is there another inertial frame in which the fields $\vec{E}''$ and $\vec{B}''$ are perpendicular to each other?*

(You may assume without loss of generality that the inertial frames are in relative motion parallel to the x-axis, and that neither $\vec{B}$ nor $\vec{E}$ has an x-component. Why? Briefly justify your assumptions.)

(a) Yes. Setting the cross product of Equations (11.46) and (11.45) equal to zero, and using the vector identity $\vec{u} \times (\vec{v} \times \vec{w}) = (\vec{u} \cdot \vec{w})\vec{v} - (\vec{u} \cdot \vec{v})\vec{w}$ and the fact that $\vec{v}$ is perpendicular to both $\vec{E}$ and $\vec{B}$, leads to

$$\frac{\frac{\vec{v}}{c}}{1 + \frac{|\vec{v}|^2}{c^2}} = \frac{\vec{E} \times c\vec{B}}{|\vec{E}|^2 + c^2|\vec{B}|^2}. \tag{12.3}$$

(b) No. According to Equation (11.93), $\vec{E} \cdot \vec{B}$ is invariant.

➤ Chapter 13 ≺

Beyond Special Relativity

Next Stop: General Relativity!

13.1 Problems with Special Relativity

We began in Chapter 2 by using moving trains to model inertial reference frames. But we made an implicit assumption beyond assuming an ideal train, with no friction and a perfectly straight track. We also assumed that there was no gravity. Einstein's famous thought experiment for discussing gravity is to consider a "freely falling" reference frame, typically a falling elevator. Objects thrown horizontally in such an elevator will not be seen to fall—there is no gravity (for a little while, at least). But even here there is an implicit assumption, namely, that the elevator is small compared to the earth.

Returning to our ideal train moving to the right, now assume there is gravity. A ball thrown straight up will, according to an observer on the train, eventually turn around and fall back down, moving along a straight line, as shown in Figure 13.1(a). Since the acceleration due to gravity at the surface of the earth is taken to be constant, the exact motion is described by a quadratic equation in t. From the ground, the same effect is seen, but combined with a constant motion (linear in t) to the right. The motion therefore takes place along a parabola; see Figure 13.2.

Now suppose that the train is also accelerating to the right with constant acceleration. Then a ball thrown straight up on the train still moves along the same parabolic trajectory as before (as seen from the ground), but the rear wall of the train might now catch up with it before it lands. This shows, first of all, that Newton's laws fail in a noninertial frame.

But analyze the situation more carefully. Try to compensate by giving the ball's velocity a horizontal component. If the ball's initial horizontal

FIGURE 13.1. A ball thrown (a) straight up in a train moving at constant speed and (b) at an angle in a train moving with constant acceleration, as seen from the train.

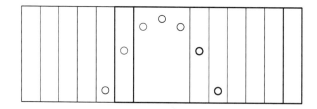

FIGURE 13.2. A ball thrown in a moving train, as seen from the ground.

speed is chosen appropriately, the resulting trajectory could look like the "boomerang" in Figure 13.1(b), in which (as seen from the train) the ball is thrown up at an angle and returns along the same path.

What is going on here? Acceleration and gravity produce the same kind of effect, and what the "boomerang" is telling us is that the *effective* force of gravity is no longer straight down, but rather at an angle. If we throw the ball up at that angle, it goes "up" and "down" in a straight line.

13.2 TIDAL EFFECTS

Consider now two objects falling toward the earth, but far from it, as shown in Figure 13.3(a). Both objects fall toward the center of the earth—which is not quite the same direction for each object. Assuming the objects start at the same distance from the earth, their paths will converge. Now, if they don't realize they are falling—by virtue of being in a large falling elevator,

13.2. TIDAL EFFECTS

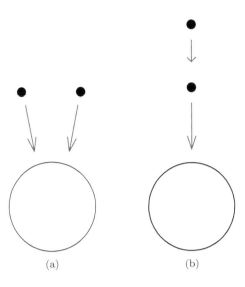

FIGURE 13.3. Two objects falling toward the earth from far away either move (a) closer together or (b) farther apart, depending on their initial configuration.

say—they will nevertheless notice that they are approaching each other. This is gravity.

Similarly, if, as shown in Figure 13.3(b), one object starts out above the other, it is slightly farther from the earth, and hence experiences a slightly weaker gravitational attraction. Thus, the lower object will always accelerate more than the upper one, and so the distance between the objects will increase.

This is the effect that causes tides. To see this, imagine that the earth is falling toward the moon, as shown in Figure 13.4. The earth itself can be regarded as a rigid body, falling toward the moon with the velocity given by the arrow at its center. But the water on the surface of the earth falls toward the moon, depending on its location, with velocities given by the four arrows shown in the figure. The relative motion of the water and the earth is given by comparing the arrows: *both* the nearest and farthest points from the moon move away from the center, causing high tides, whereas the points on either side move toward the center, causing low tides. This explains why there are two, not one, high (and low) tides every day. A similar effect, but roughly half as strong, is caused by the sun.

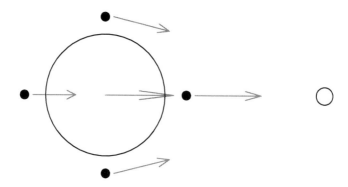

FIGURE 13.4. How tides are caused by the earth falling toward the moon.

13.3 DIFFERENTIAL GEOMETRY

In two dimensions, Euclidean geometry is the geometry of a *flat* piece of paper. But there are also *curved* two-dimensional surfaces. The simplest of these is the sphere, which has constant positive curvature and is a model of (double) *elliptic geometry*. Another important example is the hyperboloid, which also has constant curvature and is a model of *hyperbolic geometry*. Hyperbolic and elliptic geometry form the two main categories of non-Euclidean geometries.

In fact, *any* two-dimensional surface in Euclidean 3-space provides a possible geometry, most of which are curved. However, it is important to realize that distances are always positive in all such geometries. We measure the distance between two points on such a surface by stretching a string between them *along the surface*. This does not measure the (three-dimensional) Euclidean distance between the points, however; instead, it corresponds to integrating the arc length along the shortest path between them.

In hyperbola geometry, we instead made a fundamental change to the distance function, allowing it to become negative or zero. If there is precisely one (basis) direction in which distances turn out to be negative, such geometries are said to have *Lorentzian signature*, as opposed to the *Euclidean signature* of ordinary surfaces. As implied by the way we have drawn it, hyperbola geometry turns out to be flat in a well-defined sense, which immediately raises the question of whether there are *curved* geometries with Lorentzian signature.

Signature	Flat	Curved
$(+ + ... +)$	Euclidean	Riemannian
$(- + ... +)$	Minkowskian	Lorentzian

TABLE 13.5. Classification of geometries.

The mathematical study of curved surfaces forms a central part of *differential geometry*, and the further restriction to surfaces on which distances are positive is known as *Riemannian geometry*. The much more difficult case of Lorentzian signature is known, not surprisingly, as *Lorentzian geometry*, and the important special case in which the curvature vanishes is known as *Minkowskian geometry*. This classification of geometries by signature and curvature is summarized in Table 13.5.

What does this have to do with physics? We have seen that hyperbola geometry, more correctly called Minkowski space, is the geometry of special relativity. Lorentzian geometry turns out to be the geometry of general relativity. In short, according to Einstein, gravity is curvature!

13.4 General Relativity

Just as one studies flat (two-dimensional) Euclidean geometry before studying curved surfaces, one studies special relativity before general relativity. But the analogy goes much further.

The basic notion in Euclidean geometry is the distance between two points, which is given by the Pythagorean theorem. The basic notion on a curved surface is still the distance "function," but this is now a statement about infinitesimal distances. Euclidean geometry is characterized by the infinitesimal distance function

$$ds^2 = dx^2 + dy^2, \tag{13.1}$$

which can be used to find the distance along any curve. An infinitesimal distance function such as Equation (13.1) is also called a *line element* or a *metric*.

The simplest curved surface, a sphere of radius r, can be characterized by the line element

$$ds^2 = r^2\, d\theta^2 + r^2 \sin^2\theta\, d\phi^2. \tag{13.2}$$

Remarkably, it is possible to calculate the curvature of the sphere from this line element alone; it is *not* necessary to use any three-dimensional geometry. It also is possible to calculate the "straight lines"; that is, the shortest path between two given points. These are straight lines in the plane but great circles on the sphere.

Similarly, special relativity is characterized by the line element[1]

$$ds^2 = -dt^2 + dx^2, \qquad (13.3)$$

which is every bit as flat as a piece of paper. We get general relativity simply by considering more general line elements.

Of course, it's not quite that simple. The line element must have a minus sign. And the curvature must correspond to a physical source of gravity—that's where Einstein's field equations come in. But, given a line element, we can again calculate the "straight lines," which now correspond to freely falling observers. Matter curves the universe, and the curvature tells objects which paths are straight—that's gravity!

Here are two examples to whet the reader's appetite further.

First, the (three-dimensional) line element

$$ds^2 = -dt^2 + \sin^2(t) \left(d\theta^2 + \sin^2\theta \, d\phi^2 \right) \qquad (13.4)$$

describes the two-dimensional surface of a spherical balloon whose radius changes with time. This roughly corresponds to a cosmological model for an expanding universe produced by a Big Bang. Studying the properties of this model more carefully leads to good questions about relativistic cosmology.

Finally, the line element

$$ds^2 = -\left(1 - \frac{2m}{r}\right) dt^2 + \frac{dr^2}{1 - \frac{2m}{r}} \qquad (13.5)$$

describes a simplified model of a black hole, with an apparent singularity at $r = 2m$, which is, however, just due to a poor choice of coordinates. Trying to understand what actually happens at $r = 2m$ will give some understanding of what a black hole really is.

To pursue these ideas further, take a look at an introductory textbook on general relativity, such as d'Inverno [d'Inverno 92] or Taylor and Wheeler [Taylor and Wheeler 00].

---

[1] Note that we have set $c = 1$.

13.5 Uniform Acceleration and Black Holes

Special relativity is about inertial observers moving at constant velocity. Consider now an observer undergoing uniform acceleration. What does this mean?

We start with Newton's second law in the form

$$F = \frac{dp}{dt} = ma, \tag{13.6}$$

but we use the relativistic notion of momentum, so that

$$p = mc \sinh \beta. \tag{13.7}$$

Setting a = constant is equivalent to assuming that p is a linear function of t, leading to

$$\sinh \beta = \frac{at}{c}. \tag{13.8}$$

But we also have

$$v = \frac{dx}{dt} = c \tanh \beta, \tag{13.9}$$

so that

$$dx = \frac{c^2}{a} \sinh \beta \, d\beta, \tag{13.10}$$

and it is easy to solve this differential equation for x.

Trajectories of uniform acceleration are therefore given by

$$x = \frac{c^2}{a} \cosh \beta \tag{13.11}$$

and

$$t = \frac{c}{a} \sinh \beta. \tag{13.12}$$

In other words, uniform acceleration corresponds to worldlines that are hyperbolas, as shown in Figure 13.6. These hyperbolas are just our old friends—"circles" of constant distance from the origin. We now discuss the physical implications of their geometric properties.

First of all, the instantaneous speed of the accelerating object is given by the (inverse) slope of the worldline at that point. A hyperbolic worldline describes an object that is instantaneously at rest at $t = 0$, but whose speed approaches the speed of light as t goes to ∞. No, we can't actually travel at the speed of light, but with uniform acceleration we can come arbitrarily close.

The asymptotes of our hyperbolic worldline are lines at 45°, and could therefore represent beams of light. This means that an object undergoing

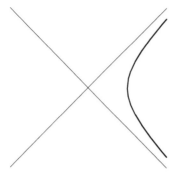

FIGURE 13.6. The trajectory of a uniformly accelerating object.

uniform acceleration can outrun a light beam. All that is necessary is to have a head start—and to end the race after finite time.

Furthermore, the region of spacetime behind that light beam is invisible to the uniformly accelerating object. More precisely, no signal originating at $x < 0$ when $t = 0$ can ever reach the object. No light can escape this region. This is precisely the geometric idea behind black holes.

The geometry shown in Figure 13.6 is, in fact, very similar to that of the Schwarzschild black hole. "Stationary" observers far away remain at a constant distance from the black hole, despite the gravitational field; they are therefore undergoing uniform acceleration equal and opposite to that of gravity. A region of spacetime exists, "inside" the black hole, from which a light beam can never escape. Both of these properties are correctly modeled by our much simpler model involving uniform acceleration but no gravity.

⋟ CHAPTER 14 ⋞

HYPERBOLIC GEOMETRY

In which the relationship between hyperbola geometry and hyperbolic geometry is clarified.

14.1 NON-EUCLIDEAN GEOMETRY

In two dimensions, *Euclidean* geometry is the geometry of an infinite sheet of paper. The postulates now exist in several different forms, but all address the basic properties of lines and angles. Key among them is the *parallel postulate*, which states: given any line and a point not on that line, there exists a unique line through the given point that is parallel to the given line.

It was thought by Euclid (and many after him) that the parallel postulate was so obvious that it should follow from the other postulates. In the centuries following Euclid, many incorrect proofs were proposed, purporting to show just that, but the claim is false; the parallel postulate turns out to be independent of the others. These attempted proofs ultimately gave rise in the early nineteenth century to hyperbolic geometry, the geometry in which the parallel postulate fails due to the existence of more than one parallel line.

The existence of such non-Euclidean geometries was ultimately established by considering particular models of hyperbolic geometry, several of which we discuss in this chapter. However, just as the sphere provides the most intuitive model for *elliptic* geometry, in which there are no parallel lines, the hyperboloid in Minkowski space provides the most intuitive model of hyperbolic geometry. But Minkowski space and special relativity did not exist until the early twentieth century, and were therefore not available when these other models were being developed.

In this chapter, we run history backward: we first describe the hyperboloid model of hyperbolic geometry, and we then discuss briefly how to derive the most common traditional models of hyperbolic geometry from the hyperboloid.

14.2 THE HYPERBOLOID

We move now to three-dimensional Minkowski space, with the squared distance function given infinitesimally (with $c = 1$) by

$$ds^2 = dx^2 + dy^2 - dt^2, \tag{14.1}$$

which can, of course, be positive, zero, or negative. We consider the hyperboloid

$$x^2 + y^2 - t^2 = -\rho^2, \tag{14.2}$$

where the radius ρ is a positive constant, and we further restrict ourselves to the branch with $t > 0$. This is the hyperboloid shown in Figure 14.1.

We emphasize that, just as with the sphere $x^2 + y^2 + z^2 = \rho^2$, the hyperboloid is a *two*-dimensional model, even though it lives in three dimensions because there are only two independent directions at any point *on* the hyperboloid.

To turn the hyperboloid into a geometric model, we must define its points and lines. Points are, well, points, but what should we count as "straight lines," henceforth simply called *lines*? Again, we turn to the sphere for guidance.

The lines on a sphere are the great circles, namely, all possible equators of the sphere. Be careful here; lines of constant longitude are great circles, but lines of constant latitude are not. Lines of constant latitude are not straight.[1] But what corresponds to an "equator" on the hyperboloid? A better characterization of lines (great circles) on the sphere is that they are the intersections of the sphere with all possible planes through the origin. This definition carries over unchanged to the hyperboloid.

How do we find the distance between two points? On the sphere, it's easy: find the angle between the corresponding vectors, and multiply by the radius. But how do we find the angle? Use the dot product. This procedure works just as well on the hyperboloid.

---

[1] Among other things, this fact provides a geometric explanation of why a Foucault pendulum changes its apparent plane of motion as the earth rotates.

14.2. THE HYPERBOLOID

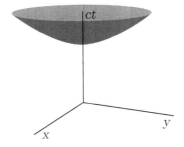

FIGURE 14.1. The hyperboloid in three-dimensional Minkowski space.

Recall the formula (5.31) for the dot product of two timelike vectors. If $\vec{u}$ and $\vec{v}$ are vectors from the origin to points U and V, respectively, on our hyperboloid, then the distance between U and V is just

$$d_H(U,V) = -\rho \cosh^{-1}\left(\frac{\vec{u} \cdot \vec{v}}{\rho^2}\right). \quad (14.3)$$

This situation is shown in Figure 14.2.

We can verify this formula in simple cases by using coordinates. Introducing natural "hyperboloidal" coordinates, we can write

$$U = (\rho \sinh \beta \cos \phi, \rho \sinh \beta \sin \phi, \rho \cosh \beta), \quad (14.4)$$
$$V = (\rho \sinh \alpha \cos \psi, \rho \sinh \alpha \sin \psi, \rho \cosh \alpha), \quad (14.5)$$

and if $\phi = 0 = \psi$, we are back in two dimensions—that is, in hyperbola geometry. In this case, the distance between U and V is just

$$\begin{aligned} d_H(U,V) &= \rho \cosh^{-1}(-\sinh \alpha \sinh \beta + \cosh \alpha \cosh \beta) \\ &= \rho \cosh^{-1}(\cosh(\beta - \alpha)) \quad (14.6) \\ &= \rho(\beta - \alpha), \end{aligned}$$

as expected.

Infinitesimal distance on the hyperboloid is obtained by rewriting Equation (14.1) for infinitesimal distance in terms of hyperboloidal coordinates (ρ,β,ϕ). We obtain

$$ds^2 = \rho^2 \, d\beta^2 + \rho^2 \sinh^2 \beta \, d\phi^2 - d\rho^2, \quad (14.7)$$

which is not at all surprising if we recall the similar computation in spherical coordinates. Since $\rho =$ constant on the hyperboloid, infinitesimal distance

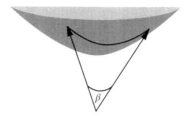

FIGURE 14.2. Using the hyperbolic dot product to measure distance on the hyperboloid.

on the hyperboloid is given by

$$ds^2 = \rho^2\, d\beta^2 + \rho^2\, \sinh^2 \beta\, d\phi^2, \tag{14.8}$$

which again recalls the line element on a sphere.

It is important to note that even though we use Minkowski space in our construction, with its indefinite distance function, the end result is a positive-definite (Riemannian) notion of distance. There are no light beams within the hyperboloid.

14.3 THE POINCARÉ DISK

The Poincaré disk model of hyperbolic geometry is obtained from the hyperboloid by stereographic projection. Stereographic projection of an ordinary sphere maps the sphere to the xy-plane by projecting from the south pole; points in the northern hemisphere ($z > 0$) map to the interior of a disk, points in the southern hemisphere ($z < 0$) map to its exterior, and the equator ($z = 0$) maps to the circle bounding the disk.

The idea of stereographic projection is the same for our Minkowskian hyperboloid. But what are the poles? As before, they are the two points where the t-axis intersects the hyperboloid, but now we need both branches. So the south pole is the point

$$P = (0, 0, -\rho) \tag{14.9}$$

(just as for the sphere), and stereographic projection maps each point on the $t > 0$ branch of the hyperboloid to the interior of a disk in the xy-plane.

Stereographic projection is depicted in Figure 14.3, which shows a triangle in a vertical slice through the t-axis, taken for simplicity to be in the

14.3. THE POINCARÉ DISK

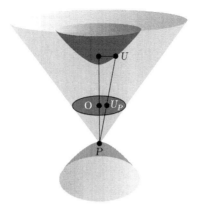

FIGURE 14.3. Stereographic projection of the hyperboloid.

xt-plane. The coordinates of the point U are then

$$U = (\rho \sinh \beta, 0, \rho \cosh \beta), \qquad (14.10)$$

and the use of similar triangles tells us that the image point U_P has coordinates

$$U_P = \left(\frac{\rho \sinh \beta}{1 + \cosh \beta}, 0, 0\right). \qquad (14.11)$$

The image of the general point U, with coordinates as given by (14.4), would be

$$U_P = (R \cos \phi, R \sin \phi, 0), \qquad (14.12)$$

where

$$R = \frac{\rho \sinh \beta}{1 + \cosh \beta} \qquad (14.13)$$

is the radial coordinate in the Poincaré disk. It is now a straightforward exercise in algebra to verify that the line element on the hyperboloid, rewritten in terms of R, is just

$$ds^2 = \frac{4\rho^4 \left(dR^2 + R^2 \, d\phi^2\right)}{(\rho^2 - R^2)^2}, \qquad (14.14)$$

in agreement with the literature for the line element of the Poincaré disk (which normally assumes $\rho = 1$).

We have therefore shown by direct computation that the Poincaré disk model can be thought of as a relabeling of the points on the hyperboloid;

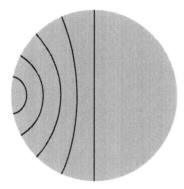

FIGURE 14.4. Some hyperbolic lines in the Poincaré disk.

these models are completely equivalent. The Poincaré disk has the twin advantages of living in two dimensions and not requiring Minkowski space for its construction, but the hyperboloid has the advantage of sharing many obvious symmetries with the sphere.

So what is the Poincaré disk *model*? As before, a geometric model is specified by giving its points and lines. Points are points, namely the interior of the unit disk in the xy-plane. But what are the lines? Since the lines on the hyperboloid are contained in planes through the origin, but stereographic projection uses lines from the south pole, the answer to this question is not immediately obvious. It turns out that *lines* in the Poincaré disk model are arcs of *Euclidean* circles that intersect the boundary of the disk orthogonally (at both points of intersection). A special case is, of course, radial lines, corresponding to circles of infinite radius. Some hyperbolic lines are shown in Figure 14.4.

A closely related model is the *Poincaré half-plane*, which is obtained from the Poincaré disk by using a complex fractional linear transformation, namely,

$$X + iY = 2\frac{i(x+iy) + \rho}{(x+iy) + \rho i}, \tag{14.15}$$

where $x + iy = Re^{i\phi}$. Another exercise in algebra verifies that this transformation maps the unit disk onto the half-plane $y \geq 0$, and further that

$$ds^2 = \frac{4\rho^4 \left(dx^2 + dy^2\right)}{(1 - x^2 - y^2)^2} = \rho^2 \frac{dX^2 + dY^2}{Y^2} \tag{14.16}$$

(where $x^2 + y^2 = R^2$), which again agrees with the literature (for $\rho = 1$).

The points in the Poincaré half-plane model of hyperbolic geometry are the points in the xy-plane with $y > 0$. The lines are open, Euclidean semicircles of any radius centered at points along the x-axis, including vertical lines, which correspond to an infinite Euclidean radius.

Both the Poincaré disk and Poincaré half-plane models are *conformal* models in that their line elements are multiples of the ordinary Euclidean metric. Since all distances are merely rescaled, angles can be defined through the dot product and are unaffected by the rescaling, which cancels out. Thus, angles in both of these models can be measured with Euclidean protractors.

14.4 THE KLEIN DISK

The Klein disk model of hyperbolic geometry is also a projection of the hyperboloid. This time, however, we project from the origin into the plane $t = \sigma$. Thus, the point U, with coordinates as given by (14.4), would be mapped to the point

$$U_K = (\sigma \tanh \beta \cos \phi, \sigma \tanh \beta \sin \phi, \sigma), \qquad (14.17)$$

as shown in Figure 14.5. The image of the hyperboloid is again the interior of a disk—this time, the disk of radius σ in the plane $t = \sigma$. The constant σ can lie between 0 and ρ, as shown in the figure, although it is useful to choose $\sigma = \rho$, in which case the disk is actually tangent to the hyperboloid at its tip.

Since lines on the hyperboloid are contained in planes through the origin, and since planes intersect in straight (Euclidean) lines, the Klein disk

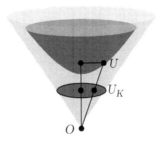

FIGURE 14.5. Constructing the Klein disk.

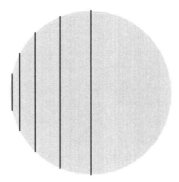

FIGURE 14.6. Some hyperbolic lines in the Klein disk.

model has the nice feature that lines are Euclidean; the lines of the Klein disk model are the chords of the circle bounding the disk, as shown in Figure 14.6. This feature is at first sight much nicer than any of the previous models, in which "lines" bear little resemblance to Euclidean lines. However, this simplicity comes at a price. More algebra reveals that the hyperboloid line element, rewritten by using coordinates from the Klein disk, takes the form

$$ds^2 = \frac{\rho^4\, dK^2}{(\rho^2 - K^2)^2} + \frac{\rho^2 K^2\, d\phi^2}{(\rho^2 - K^2)}, \tag{14.18}$$

where

$$K = \rho \tanh \beta \tag{14.19}$$

is the radial coordinate in the Klein disk. Due to the different denominators in the two terms, the line element is no longer conformal to the Euclidean metric. This means that angles *cannot* be measured with a Euclidean protractor in the Klein disk model.

14.5 THE PSEUDOSPHERE

Each of the preceding models of hyperbolic geometry discussed in this chapter has its advantages and disadvantages. The two Poincaré models are conformal, so angles are easy to measure, but lines are not obvious. Lines in the Klein model are obvious, but angles are difficult to measure. And for all its similarity with the sphere, in the hyperboloid model neither lines nor

14.5. THE PSEUDOSPHERE

angles are easy to determine. Furthermore, none of these models exists in ordinary three-dimensional Euclidean space. Perhaps that is no surprise; after all, why should a model for hyperbolic geometry be constructible within Euclidean geometry? The surprise is that this is indeed possible, as we now show.

The pseudosphere is most simply related to the Poincaré half-plane, which we described in terms of dimensionless variables (X,Y) in Section 14.3. We now introduce new variables α and θ via

$$\sin\alpha = \frac{1}{Y}, \qquad (14.20)$$

$$\theta = X, \qquad (14.21)$$

and compute

$$ds^2 = \rho^2 \frac{dX^2 + dY^2}{Y^2} = \rho^2 \left(\cot^2\alpha\, d\alpha^2 + \sin^2\alpha\, d\theta^2\right). \qquad (14.22)$$

We claim that this line element describes a surface of revolution in Euclidean space.

A surface of revolution in ordinary three-dimensional Euclidean space can be described in cylindrical coordinates (r,θ,z) by specifying r and z in terms of some parameter α. The line element on the surface is then

$$ds^2 = dr^2 + r^2\, d\theta^2 + dz^2 = \left(\left(\frac{dr}{d\alpha}\right)^2 + \left(\frac{dz}{d\alpha}\right)^2\right) d\alpha^2 + r^2\, d\theta^2. \qquad (14.23)$$

Comparing Equations (14.23) and (14.22), we must have

$$r = \rho \sin\alpha, \qquad (14.24)$$

and therefore,

$$\left(\frac{dz}{d\alpha}\right)^2 = \rho^2 \cot^2\alpha - \rho^2 \cos^2\alpha = \rho^2 \frac{\cos^4\alpha}{\sin^2\alpha}. \qquad (14.25)$$

One solution of this differential equation is

$$z = -\rho\cos\alpha - \rho\ln\tan\left(\frac{\alpha}{2}\right). \qquad (14.26)$$

Equations (14.24) and (14.26) with $\alpha \in [0, \pi/2]$ describe a curve known as a *tractrix*, as shown in Figure 14.7(a). The *pseudosphere* is the surface of revolution obtained by rotating the tractrix about the z-axis, as shown in Figure 14.7(b). We can also embed the pseudosphere in one of the

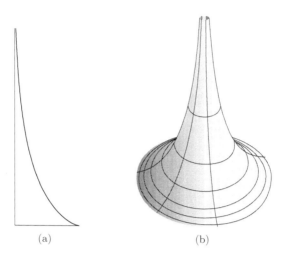

FIGURE 14.7. (a) The tractrix, whose surface of revolution is (b) the pseudosphere.

other models of hyperbolic geometry, as shown for the hyperboloid in Figure 14.8. The periodicity in θ requires that we identify the edges as shown in the figure; we can think of the pseudosphere as a "rolled-up" piece of the hyperboloid. We can then think of the horizontal slices of the pseudosphere as circles centered at infinity (the tip of the pseudosphere), with infinite radius (the length from the center, as measured along the pseudosphere).

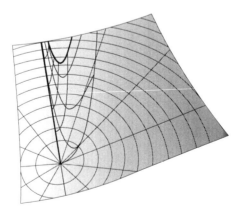

FIGURE 14.8. Embedding the pseudosphere in the hyperboloid. The heavy lines are to be identified.

14.5. The Pseudosphere

But what is so special about the pseudosphere?

A Euclidean plane has the property that it is the same everywhere, and in all directions. Thus, we can place a piece of cardboard on a flat table anywhere, and in any orientation, and it will fit snugly. We describe this by saying that the table is flat (as is the piece of cardboard); its curvature is zero. A sphere has the same property. For example, if we peel an orange, we can put a piece of the peel back anywhere on the orange, in any orientation, and it will fit snugly. The orange has constant (positive) curvature. The hyperboloid also has this property; it has constant negative curvature. If you could peel it, the peel would fit snugly anywhere, and in any orientation. But the hyperboloid lives in Minkowski space, so this property seems to be far removed from reality.

Yet we have succeeded in building a model of the hyperboloid in Euclidean space. This model has exactly the same intrinsic properties as any other model of hyperbolic geometry, including the hyperboloid. This means that if we were to build a pseudosphere, and then "peel" it,[2] we could take that piece of peel and put it anywhere, in any orientation, and it would fit snugly, without tearing or stretching.

This is indeed remarkable.

---

[2] One way to do this would be to use the pseudosphere as a mold to make a partial copy, for instance with a spray-on rubber-like product. Robert Osserman, a well-known geometer, has done this demonstration during public lectures.

CHAPTER 15

CALCULUS

In which a simple derivation of the derivatives of trigonometric and exponential functions is presented, without the use of limits, numerical estimates, solutions of differential equations, or integration.

15.1 CIRCLE TRIGONOMETRY

Consider once again the point P on a circle of radius r with coordinates $(r\cos\theta, r\sin\theta)$, as shown in Figure 3.2, and recall the geometric definitions of the basic trigonometric functions (3.2) and (3.3).

We would like to compute the derivatives of these functions. What do we know? We know that (infinitesimal) arc length along the circle is given by

$$ds = r\, d\theta, \qquad (15.1)$$

but we also have the (infinitesimal) Pythagorean theorem, which tells us that

$$ds^2 = dx^2 + dy^2. \qquad (15.2)$$

Furthermore, from $x^2 + y^2 = r^2$, we obtain

$$x\, dx + y\, dy = 0. \qquad (15.3)$$

Putting this information together, we have

$$r^2\, d\theta^2 = dx^2 + dy^2 = dx^2\left(1 + \frac{x^2}{y^2}\right) = r^2\, \frac{dx^2}{y^2}, \qquad (15.4)$$

so that

$$d\theta^2 = \frac{dx^2}{y^2} = \frac{dy^2}{x^2}, \qquad (15.5)$$

The material in this chapter is based on the presentation in [Dray 13].

where we have used Equation (15.3) in the last step. By carefully using Figure 3.2 to check signs, we can take the square root of Equation (15.5) and rearrange terms to obtain

$$dy = x\, d\theta,$$
$$dx = -y\, d\theta. \tag{15.6}$$

Finally, by inserting definitions (3.2) and (3.3) and using the fact that $r = \text{constant}$, we recover the familiar expressions

$$d\sin\theta = \cos\theta\, d\theta,$$
$$d\cos\theta = -\sin\theta\, d\theta. \tag{15.7}$$

We have thus determined the derivatives of the basic trigonometric functions from little more than the geometric definition of those functions and the Pythagorean theorem—and the ability to differentiate simple polynomials.

15.2 Hyperbolic Trigonometry

The derivation in Section 15.1 carries over virtually unchanged to hyperbolic trigonometric functions. Recall from Section 4.3 the geometric definitions of the hyperbolic trigonometric functions (4.13) and (4.14), as shown in Figure 4.2.

We compute the derivatives of these functions by using the same technique as before. We choose to work on the hyperbola $y^2 - x^2 = \rho^2$, which is the hyperbola shown in Figure 4.2 that contains the point B. What do we know? We know that the (infinitesimal) arc length along the hyperbola is given by

$$ds = \rho\, d\beta, \tag{15.8}$$

but we also have the (infinitesimal, Lorentzian) Pythagorean theorem, which tells us that

$$ds^2 = dx^2 - dy^2. \tag{15.9}$$

Furthermore, from $y^2 - x^2 = \rho^2$, we obtain

$$x\, dx = y\, dy. \tag{15.10}$$

Putting this information together, we have

$$\rho^2\, d\beta^2 = dx^2 - dy^2 = dx^2\left(1 - \frac{x^2}{y^2}\right) = \rho^2\, \frac{dx^2}{y^2}, \tag{15.11}$$

15.3. EXPONENTIALS (AND LOGARITHMS)

so that
$$d\beta^2 = \frac{dx^2}{y^2} = \frac{dy^2}{x^2}, \tag{15.12}$$

where we have used Equation (15.10) in the last step. By carefully using Figure 4.2 to check signs, we can take the square root of Equation (15.12) and rearrange terms to obtain
$$dy = x \, d\beta,$$
$$dx = y \, d\beta. \tag{15.13}$$

Finally, by inserting definitions (4.13) and (4.14) and using the fact that $\rho = $ constant, we obtain
$$d \sinh \beta = \cosh \beta \, d\beta,$$
$$d \cosh \beta = \sinh \beta \, d\beta. \tag{15.14}$$

We have thus determined the derivatives of the basic hyperbolic trigonometric functions.

15.3 EXPONENTIALS (AND LOGARITHMS)

Hyperbolic trigonometric functions are usually defined by using formulas (4.1) and (4.2), and it takes some work (and independent knowledge of the exponential function) to show directly that our definition is equivalent to this one. We turn this on its head and instead *define*
$$\exp(\beta) = \cosh \beta + \sinh \beta. \tag{15.15}$$

The function $\exp(\beta)$ as defined in (15.15) has an immediate geometric interpretation, as shown in Figure 15.1, where it is important to recall that β is the *hyperbolic* angle, and not the Euclidean angle that would be measured by a (Euclidean) protractor.

We immediately have
$$d\left(\exp(\beta)\right) = \sinh \beta \, d\beta + \cosh \beta \, d\beta = \exp(\beta) \, d\beta, \tag{15.16}$$

but it remains to show that $\exp(\beta)$ as defined above is really the same as e^β. It is enough, however, to show that $\exp(\beta)$ is *an* exponential function, that is, that it satisfies
$$\exp(0) = 1, \tag{15.17}$$
$$\exp(u+v) = \exp(u) \exp(v), \tag{15.18}$$

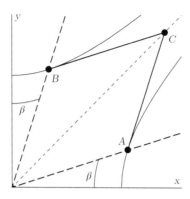

FIGURE 15.1. The geometric definition of $\exp(\beta)$. Points A and B are as in Figure 4.2, so that the coordinates of point C are $(\rho \exp(\beta), \rho \exp(\beta))$.

and to then *define*

$$e = \exp(1). \tag{15.19}$$

We can then verify that this definition of e agrees with the usual one in several ways, such as by resorting to uniqueness results for the solutions of differential equations.

First, however, we need to verify conditions (15.17) and (15.18). The first of these, (15.17), follows immediately from the fact that $\beta = 0$ corresponds to putting point A at $(\rho, 0)$ in Figure 4.2. Examining condition (15.18), we see that we need to know the addition formulas for the hyperbolic trigonometric functions, which we now construct.[1]

We first construct right triangles whose legs are not horizontal and vertical. We recall that the derivative of a vector of constant magnitude must be perpendicular to that vector, since

$$\vec{v} \cdot \vec{v} = \text{constant} \implies 0 = d(\vec{v} \cdot \vec{v}) = 2\, \vec{v} \cdot d\vec{v}. \tag{15.20}$$

Since the position vector to points on a circle centered at the origin has constant magnitude, we have shown that radii are orthogonal to circles. This argument holds just as well in hyperbola geometry, and establishes orthogonality between the radial lines intersecting our hyperbolas and the tangent lines to the hyperbolas at the point of intersection.

Consider now the situation shown in Figure 15.2. In Figure 15.2(a), angle α is in standard position, leading to the right triangle shown, with

---

[1] As in the Euclidean case, an argument that uses the geometric and algebraic properties of the Lorentzian dot product is possible, but it is less intuitive.

15.3. EXPONENTIALS (AND LOGARITHMS)

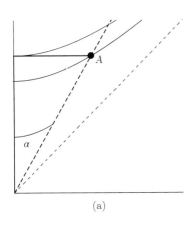
(a)

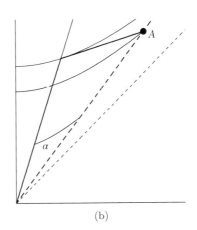
(b)

FIGURE 15.2. Using a shifted right triangle to represent the hyperbolic trigonometric functions. (a) Angle α is in standard position. (b) Angle α is shifted to the right.

radial leg $\rho \cosh \alpha$ and horizontal leg $\rho \sinh \alpha$. In Figure 15.2(b), angle α has been shifted to the right, but since the hypotenuse and radial leg of the resulting right triangle have the same length as before, so must the remaining leg; the triangles in the two drawings are congruent in hyperbola geometry.

Redraw Figure 15.2(b) as shown in Figure 15.3, so that point B corresponds to angle β, and point A corresponds to angle $\alpha + \beta$ (as measured from the y-axis). Thus, the coordinates of these points are given by

$$B = (\rho \sinh \beta, \rho \cosh \beta),$$
$$A = (\rho \sinh(\alpha + \beta), \rho \cosh(\alpha + \beta)). \quad (15.21)$$

Introduce point C as shown in Figure 15.3, so that the distance from the origin to C is $\rho \cosh \beta$; that is, so that the coordinates of C are given by

$$C = B \cosh \alpha. \quad (15.22)$$

But by the discussion of Figure 15.2, the other leg of the triangle has length $\rho \sinh \beta$, so we must also have

$$C = A - B^\perp \sinh \alpha, \quad (15.23)$$

where

$$B^\perp = (\rho \cosh \beta, \rho \sinh \beta) = \frac{dA}{d\beta} \quad (15.24)$$

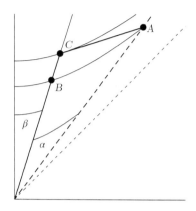

FIGURE 15.3. The geometric construction of the hyperbolic addition formulas.

denotes the direction tangent to the hyperbola at B. By combining Equations (15.22) and (15.23), we find that

$$A = B \cosh\alpha + B^\perp \sinh\alpha, \tag{15.25}$$

and we have established the addition formulas[2]

$$\begin{aligned}\sinh(\alpha+\beta) &= \sinh\alpha\cosh\beta + \cosh\alpha\sinh\beta, \\ \cosh(\alpha+\beta) &= \cosh\alpha\cosh\beta + \sinh\alpha\sinh\beta.\end{aligned} \tag{15.26}$$

Returning to Equation (15.18), we finally have

$$\begin{aligned}\exp(u+v) &= \cosh(u+v) + \sinh(u+v) \\ &= \cosh(u)\cosh(v) + \sinh(u)\sinh(v) \\ &\quad + \sinh(u)\cosh(v) + \cosh(u)\sinh(v) \\ &= \Big(\cosh(u) + \sinh(u)\Big)\Big(\cosh(v) + \sinh(v)\Big) \\ &= \exp(u)\exp(v),\end{aligned} \tag{15.27}$$

as desired.

A direct geometric verification is also possible, based on the construction in Figure 15.1, and as shown in more detail in Figure 15.4. Denoting the

---

[2] The reader is encouraged to work out the analogous argument for ordinary trigonometry, especially if the argument about "shifted" triangles seems unclear.

15.3. Exponentials (and Logarithms)

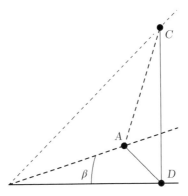

FIGURE 15.4. The geometric verification that "exp" is exponential.

origin by O, and noting that line AD is at $45°$, we see that the ratio of the length of OD to that of OA is precisely $\exp(\beta)$. In this sense, "rotation" through β (taking A to D) corresponds to stretching by a factor of $\exp(\beta)$. Composing two such "rotations" leads directly to condition (15.18); the details are left to the reader.

Having verified the desired properties of the exponential function, it is now straightforward to define logarithms as the inverse of exponentiation, that is, to define $\log(u)$ by

$$v = \log(u) \iff u = \exp(v) \tag{15.28}$$

and to establish that this definition leads to the usual properties of the natural logarithm.

Bibliography

[d'Inverno 92] Ray A. d'Inverno. *Introducing Einstein's Relativity.* Oxford: Oxford University Press, 1992.

[Dray 90] Tevian Dray. "The Twin Paradox Revisited." *Am. J. Phys.* 58 (1990), 822–825.

[Dray 03] Tevian Dray. "The Geometry of Special Relativity." Course notes, Oregon State University, 2003.

[Dray 04] Tevian Dray. "The Geometry of Special Relativity." *Physics Teacher (India)* 46 (2004), 144–150.

[Dray 10] Tevian Dray. "The Geometry of Special Relativity." Available at http://physics.oregonstate.edu/coursewikis/GSR, 2010.

[Dray 13] Tevian Dray. "Using Differentials to Differentiate Trigonometric and Exponential Functions." *College Math. J.* (to appear). Available at http://www.math.oregonstate.edu/bridge/papers/trig.pdf.

[Dray and Manogue 06] Tevian Dray and Corinne A. Manogue. "The Geometry of the Dot and Cross Products." *JOMA* 6 (June 2006). Available at http://www.joma.org/?pa=content&sa=viewDocument&nodeId=1156.

[Einstein 05] Albert Einstein. "Zur Elektrodynamik bewegter Körper." *Annalen der Physik* 322 (1905), 891–921. Available in German at http://wikilivres.info/wiki/Zur_Elektrodynamik_bewegter_K%F6rper. Available in English at http://en.wikisource.org/wiki/On_the_Electrodynamics_of_Moving_Bodies.

[Griffiths 99] David J. Griffiths. *Introduction to Electrodynamics*, Third edition. Upper Saddle River, NJ: Prentice Hall, 1999.

[Norton 06] John D. Norton. "Discovering the Relativity of Simultaneity: How Did Einstein Take 'The Step.'" In *Einstein in a Trans-cultural Perspective*, edited by Yang Jiang and Liu Bing. Beijing: Tsinghua University Press, 2006. Available at http://www.pitt.edu/~jdnorton/papers/AE_1905_Tsinghua.pdf, or online with animated graphics at http://www.pitt.edu/~jdnorton/Goodies/rel_of_sim.

[Taylor and Wheeler 63] Edwin F. Taylor and John Archibald Wheeler. *Spacetime Physics*. San Francisco: W H Freeman, 1963.

[Taylor and Wheeler 92] Edwin F. Taylor and John Archibald Wheeler. *Spacetime Physics*, Second edition. New York: W H Freeman, 1992.

[Taylor and Wheeler 00] Edwin F. Taylor and John Archibald Wheeler. *Exploring Black Holes: Introduction to General Relativity*. San Francisco: Addison-Wesley, 2000.

INDEX

acceleration, 99
addition formula
    Einstein, 11, 24, 39, 68
    for slopes, 18
    for velocities, 39
Ampère's law, 92
angles, transformation of, 52–55

Big Bang, 104
black hole, 104, 106

capacitor, 86
causality, 30
charge density, 92
circle
    in Euclidean geometry, 13
    in hyperbola geometry, 20
clairvoyante, 49
collision
    identical particles, 76–77, 80
    lumps of clay, 75–76
conservation
    of 2-momentum, 72
    of energy, 69, 72
    of mass, 69, 72
    of momentum, 69, 72
Coriolis effect, 2

cosmology, 104
current density, 92, 93
curvature, 103, 104, 117

distance, 22
    invariant, 26
    squared, 13, 20, 22, 29
distance function, *see* distance, squared
Doppler shift, 44, 59
dot product, 30–35

Einstein summation convention, 89
Einstein, Albert, 5
electric field
    of an infinite sheet, 86
    of an infinite wire, 83
electromagnetic fields
    Lorentz transformation of, 87–89
    parallel, 98
    perpendicular, 98
    scalar invariants of, 95
    tensor description of, 92
    vanishing, 97–98
energy, 71
    $E = mc^2$, 71
    conservation of, *see* conservation

ether, 5
events, 37
expanding universe, 104
exponential function
    addition formula for, 124–125
    definition of, 121
    derivative of, 121
    geometry of, 121
    properties of, 121

Faraday's equation, 92
Foucault pendulum, 2, 108

Galilean relativity, 4
Gauss's law, 92
geometric model, 108
geometry
    conformal, 113
    differential, 103
    elliptic, 102, 107
    Euclidean, 102, 107
    hyperbola, 20
    hyperbolic, 102, 107
    Lorentzian, 103
    Minkowskian, 103
    non-Euclidean, 107
    of trigonometry, 13
    Riemannian, 103
getaway, 51
gravity, 99, 101
great circles, 108

hyperboloid, 108
    infinitesimal distance on, 110

inertial frame, 2, 4
interstellar travel, 55–57
interval, 12, 29

Klein disk, 113–114

latitude, 108
length contraction, 10, 40
lightlike, 29
lightyear, 26
line element, 103
logarithms, 125

longitude, 108
Lorentz force law, 84
Lorentz transformation, 10, 27–29, 68,
    89, *see also* rotations,
    Lorentzian
Lorentz, Hendrik, 10

magnetic field
    of an infinite sheet, 86
    of an infinite wire, 83
massless particles, 72
Maxwell's equations, 4–5, 92
    tensor form of, 94
merry-go-round, 2
meter stick
    at rest, 39–40
    moving, 41
metric, 103
Michelson, Albert Abraham, 5
Michelson-Morley experiment, 5
Minkowski space, 103
    three-dimensional, 108
momentum
    2-momentum, 71
    conservation of, *see* conservation
momentum diagram, 76
moving clock, 42–43
muons
    disintegrating, 47
    from cosmic rays, 57–58

neutrino, 81
Newton's laws, 1, 99
north
    geographic, 25
    magnetic, 25

observation, 3
observer, 3
orthogonal, 38
orthonormal basis, 31
Osserman, Robert, 117

paradoxes
    manhole cover, 65
    pole and barn, 61–62
    twin, 63–65

parallel postulate, 107
permeability constant, 83
permittivity constant, 83
photons, 72, 78
pion decay, 77–79, 81–82
Planck's constant, 73
Poincaré disk, 110–112
Poincaré half-plane, 112–113
postulates, 3, 5
proper time, 67
pseudosphere, 115–117
Pythagorean theorem, 13
    hyperbolic, 23, 43
        infinitesimal, 120
    infinitesimal, 119

rapidity, 28, 38
reference frame, 3
    center-of-mass, 69, 77, 79
rest mass, 71
right angle, 31
right triangle, 32
rocket ship, 48
rotations
    Euclidean, 16
    hyperbolic, 23, 29, *see also*
        Lorentz transformation
    Lorentzian, *see* rotations,
        hyperbolic

scalar potential, 93
spacelike, 29, 37
spacetime, 26, 29–30
spacetime diagram, 26, *see also*
    momentum diagram
    3-dimensional, 42
speed of light, 5
speeding cars, 49
squared distance, *see* distance, squared
stereographic projection, 110
straight lines, 104, 108
suitcase, 17
surveyor's parable, 25

tensors, 91
    antisymmetric, 91
    Lorentz transformation of, 91
thought experiment, 5
tides, 1, 101
time dilation, 8, 42
timelike, 29, 37
tractrix, 115
train, 4–10, 42–43, 99–100
trigonometric functions
    derivative of, 119–120
    hyperbolic, *see* trigonometry,
        hyperbolic
        addition formulas for, 124
        derivative of, 120–121
trigonometry
    circular, 13–15
    hyperbolic, 19–22
    triangle, 15, 22

uniform acceleration, 105–106

vector potential, 93
vectors
    contravariant, 89
    covariant, 90
    dual, 90
    length of, 32
    lightlike, 32
    orthogonal, 31
    perpendicular, 31
    spacelike, 32
    spacetime, 31
    timelike, 32
velocity, 68
    2-velocity, 68
    4-velocity, 90

worldline, 26, 37
worldsheet, 39
wristwatch time, 67